Series / Number 07-001

ANALYSIS OF VARIANCE

GUDMUND R. IVERSEN
Swathmore College

HELMUT NORPOTH
University of Cologne

SAGE PUBLICATIONS / **Beverly Hills** / **London**

For information address:

SAGE PUBLICATIONS, INC.
275 South Beverly Drive
Beverly Hills, California 90212

SAGE PUBLICATIONS LTD
28 Banner Street
London EC1Y 8QE, England

International Standard Book Number 0-8039-0650-1

Library of Congress Catalog Card No. L.C. 76-25695

THIRD PRINTING

When citing a University Paper, please use the proper form. Remember to cite the
correct Sage University Paper series title and include the paper number. One of the
two following formats can be adapted (depending on the style manual used):

(1) IVERSEN, GUDMUND R. and NORPOTH, HELMUT (1976) "Analysis of
Variance." Sage University Paper series on Quantitative Applications in the Social
Sciences, 07-001. Beverly Hills and London: Sage Pubns.

OR

(2) Iversen, Gudmund R. and Norpoth, Helmut. 1976. *Analysis of Variance*. Sage
University Paper series on Quantitative Applications in the Social Sciences, series no.
07-001. Beverly Hills and London: Sage Publications.

CONTENTS

Editor's Introduction   5

1. Introduction   7

2. One-Way Analysis of Variance, All Categories   11
   Two Groups   11
   More Than Two Groups   25

3. Two-Way Analysis of Variance, All Categories   39
   Unrelated Explanatory Variables   39
   Related Explanatory Variables   59
   Special Topics   69

4. Analysis of Variance, Sample of Categories   73
   One-Way Analysis   73
   Two Explanatory Variables   78

5. Other Models   82
   Three Explanatory Variables   82
   Latin Square Design   83
   Nested Designs   97
   Analysis of Variance and Regression   88

6. Conclusion   91

Notes   93

References   94

Editor's Introduction

ANALYSIS OF VARIANCE is a discussion of a basic technique in the understanding of social science data. Presuming only an elementary understanding of the methods of data analysis—and the general problem of significance testing* upon which analysis of variance is based—this paper is not only valuable in its own right, but it is also useful background for reading and understanding the more advanced topics explored by other papers in this series.

ANALYSIS OF VARIANCE is one of the techniques included within the rubric of "the general linear model." Such a model assumes a linear relationship between those variables we want to predict and their possible determinants. To find out how much impact one variable has on another and to establish how good is the predictive success of a model, we use the techniques of regression analysis.**

Associated with regression analysis is ANALYSIS OF VARIANCE, which poses a different question. It does not attempt to measure the *fit* between variables. Instead, *it seeks to determine the probability that a predictor variable could yield results different from simple random selection.* This is, of course, the logic behind significance testing. ANALYSIS OF VARIANCE, then, starts with a variable to be predicted—measured on an interval or ratio scale—and one or more predictor variables grouped according to some attribute. The question then becomes:

Do the various categories for the attribute discriminate among high or low scores for the variable to be predicted?

In more concrete terms, this paper uses for illustration the assembled data on the beliefs and values of citizens in five countries (Great Britain, Italy, Mexico, the United States, and West Germany) to pose this question:

---

*Readers who wish a general introduction to this subject may consult the paper in this series by Ramon E. Henkel (1976) *Tests of Significance*. Sage University Papers on Quantitative Applications in the Social Sciences, series no. 07-004. Beverly Hills and London: Sage Publications.

**For an introduction to the techniques of regression analysis, readers may consult two papers within this series: Eric Uslaner (forthcoming) *Regression Analysis: Simultaneous Equation Estimation* and Charles Ostrom (forthcoming) *Regression Analysis: Time Series Analysis,* both in the Sage University Papers on Quantitative Applications in the Social Sciences. Beverly Hills and London: Sage Publications.

Does the country in which one lives affect the beliefs of a citizen in his own political competence?

A positive answer to this question might indicate that cultural differences are at work; or it might imply that such cultural differences are themselves a function of other variables. If there are no differences, we have to seek other variables which might account for differing attitudes about political efficacy. For example, Iversen and Norpoth create a model which considers the joint effect of both country and level of education upon citizens' beliefs, a problem of significant interest to political scientists and sociologists.

What other applications of analysis of variance are used by social scientists?

The earliest applications were primarily in agriculture and biology, but today this methodology is used in every field of science and is one of the most important statistical tools of the social sciences. It is used by:

- Political analysts, who can postulate a fixed set of groups, such as political parties, and determine the degree to which members differ in their behavior. (The reverse process allows the detection of clusters—groupings of individual observations which maximize between-group variation while minimizing variation within the group—thereby breaking down a population into maximally distinct groups.)

- Communications researchers, who can experimentally assess the impact of communication within a research setting optimally exploiting the potential of analysis of variance for causal inference.

- Economists who can contrast rates of economic growth by classifying countries into general systems and employing data sets on growth rates.

- Public policy analysts, who can evaluate programs by employing analysis of variance to assess the effect of a particular program by comparing it either to alternative programs or to the "policy" of no program at all.

- Researchers in psychology or education, who might be concerned with the effects of education and race on standard intelligence or achievement tests. In such cases, test scores would serve as the variable to be predicted, grouping race and level of education to test for differences within and between the subgroups of the sample. (Indeed analysis of variance is particularly useful in such fields as psychology and education because it can reveal potential interaction among alternative predictors, and because the predictor variables are only required to be measures at the level of nominal scales, such as race or sex.)

This list is open-ended. Most social science disciplines can add further uses for ANALYSIS OF VARIANCE. It is a technique that is relatively easy to learn and wide ranging in its applications.

—E. M. Uslaner, Series Editor

ANALYSIS OF VARIANCE

GUDMUND R. IVERSEN
Swathmore College

HELMUT NORPOTH
University of Cologne

1. INTRODUCTION

The nature of the average citizen's involvement in politics poses a question of continuing interest to the student of political life. Democratic theory postulates the citizen, among other things, to be highly interested in political affairs, knowledgeable about them, concerned over outcomes of elections, imbued with a sense of citizen duty as well as efficacy and, ultimately, active in participating in political affairs. Empirical research has attempted to assess the extent to which the average citizen approximates such standards of political behavior.

An important example of such research is Almond and Verba's study *The Civic Culture* (1963). One of its key concepts involves the notion of subjective competence, that is, a sense of competence which a citizen attributes to himself in dealing with political authorities. Specifically, this notion refers to

(a) the type of steps an individual might take,
(b) the estimated success of the proposed action, and
(c) the likelihood of actually engaging in the proposed action.

In their study Almond and Verba ascertained responses to each of those three items of subjective competence by confronting respondents with

AUTHORS' NOTE: *We are very grateful to Lawrence S. Mayer, Eric M. Uslaner, and an anonymous reader for their comments on an earlier draft.*

the hypothetical situation where a national law was being considered by one's government, which the respondent regarded as being very unjust or harmful.

Let us now take a person's subjective competence, as measured by a single score, and look at the concept from a comparative perspective. Do members of different countries differ in their level of subjective competence? One can argue that unique historical and cultural factors account for the overall level of subjective competence attained in a given country. Being unique to each country these factors would result in different overall levels of subjective competence. The research hypothesis we wish to test says that nations differ in their level of subjective competence.

Whether or not this research hypothesis is confirmed depends not only on how much nations differ in their overall level of subjective competence, but also on the degree to which members of a given country are similar in terms of subjective competence. What we need is a technique which checks the amount of variation in the scores between countries against the variation among members of the same country. Analysis of variance is such a technique.

Analysis of variance—in some ways it is a misleading name for a collection of statistical models and methods that deal with whether the means of a variable differ from one group of observations to another. While "analysis of means" may be a better name, the methods all employ ratios of variances in order to establish whether the means differ, and the name analysis of variance is here to stay. The name is often abbreviated to ANOVA, which a student in one of our classes for a long time thought was the name of an Italian statistician.

The various statistical methods that fall under this name are related to other statistical methods. For example, when we study the difference between the means for only two groups of observations, we can find out whether the two means differ significantly by computing a t-value for a difference-of-means test. This procedure is usually not thought of as being an analysis of variance, but we show in the next chapter that it is nothing but a special case of one of the simplest analysis of variance procedures. We show that by squaring the t-value one gets a ratio of two variances, and by using this ratio one has another way of telling whether there is a statistically significant difference between the two groups.

Analysis of variance methods are also related to the set of statistical methods known as regression analysis. This point is pursued further in chapter 5; here we note only that analysis of variance is usually the appropriate method when the groups of observations are created by using a nominal level variable as the independent variable in the study. In our example the nominal level variable is country, and this variable has five categories resulting in five groups of observations. Our task consists of determining whether the groups differ in their average level of the dependent variable—here, subjective competence. The dependent variable in an analysis of variance is almost always an interval level variable.

But social scientists also have groups that are formed by interval level variables instead of nominal level variables. Take education as an example. When measured as number of years completed in school all members of one group have a score of 6, all members of another group have a score of 7, and so forth. With education used as the independent variable, regression analysis is the appropriate way of studying whether the groups differ with respect to some dependent variable denoted Y; that is, whether there is a relationship between education and the dependent variable. But the differences between analysis of variance and regression models are smaller than from first glance may seem to be the case, and they can both be seen as special cases of what is known as the general linear model.

Part of the reason why regression and analysis of variance are seen as two separate sets of methods is historical. Different sciences have tended to use different statistical methods, and analysis of variance methods originally came mostly from agriculture and the work done by the late Sir Ronald A. Fisher in the period between the two world wars. He worked for many years as a statistician at an agricultural experimental station in England, and a typical question he would be called upon to investigate might be whether several different types of fertilizers give different yields or not. By using each fertilizer on several different plots of land and measuring all the yields at the end of the growing season, one would have the basis for deciding whether the fertilizers differ or not in their effectiveness.

There is a major difference between the agricultural experiment alluded to above and the question of whether countries differ in their level of subjective competence. This is not only a difference between a particular social science study and a hypothetical agricultural experiment. Instead, this is the difference between using analysis of variance methods for the type of problem for which they were originally intended, and adapting them to different types of problems. The agricultural study is an experimental study where the experimenter has complete control of the treatments to be used, while the social science study is an observational study without the possibility of such a control.

One of the key features of Fisher's contribution to what today is known as experimental design is that in the end the allocation of a particular fertilizer to a particular plot of land is a random allocation. However, people are not allocated randomly to different countries so that they can have their subjective competence scores measured years later. This lack of randomization makes it difficult to decide whether observed differences in subjective competence levels between countries can be attributed to differences in countries, or whether the differences are due to other variables that have not been brought into the analysis. For a further discussion of observational versus experimental research see Cochran (1965).

The formal theory of analysis of variance requires the observations to satisfy certain assumptions. There is the usual assumption that the observations have been collected independently of each other. Beyond that there are assumptions about certain quantities being additive and following the normal distribution. It is possible to use the data to check some of these assumptions, and we spell out the necessary assumption for each method we discuss. The theory is heavily mathematical. We feel, rather reluctantly, that the only way to fully understand a particular method is to understand the underlying mathematics. But it is also possible to get a very good working understanding of many of the methods without extensive mathematical background. We present the arguments here with a minimum of mathematics, and instead we rely on intuition and graphical presentations.

It used to be that almost any analysis of variance implied tedious work on desk calculators. Most of this work has now been taken over by computers, even though analysis of variance programs have been slow in arriving compared with other statistical programs. The computer output for analysis of variance models is sometimes difficult to understand, and the meaning of the various quantities printed by the computer is not always completely obvious. For a further discussion of this point see Francis (1973).

The presentation here deals with some of the simpler analysis of variance models. We progress from one to two independent variables, also called explanatory variables, and the resulting models are called one-way and two-way analysis of variance, respectively. We also have to distinguish between situations when we have observations on the dependent variable for all categories of the explanatory variable(s) and those when we do not. If the countries in the above example were chosen as a sample of countries from a larger population of countries, then we do not have observations for all categories of the country variable. In this case we are dealing with what is called a random model. But if we are only interested in the particular countries for which we have observations, then we do have data for all the categories of the explanatory variable. In this case we are dealing with what is called a fixed model. The contents of the various chapters is as follows:

	One explana-tory variable (one-way anova)	Two explana-tory variables (two-way anova)	Other models
All categories (fixed model)	Chapter 2	Chapter 3	Chapter 5
Sample of categories (random model)	Chapter 4	Chapter 4	

Further readings: After this elementary introduction to analysis of variance one can read further in standard social science textbooks. For much more extensive treatments on an intermediate level there are books by authors like Cochran and Cox (1957), Dunn and Clark (1974), Hays (1973), Keppel (1973), Kirk (1968), and Snedecor and Cochran (1967). The classic book on the advanced level is by Scheffé (1959).

2. ONE-WAY ANALYSIS OF VARIANCE, ALL CATEGORIES

Two Groups

A t-test for the difference between two means. The fundamental ideas of analysis of variance can be better understood if we first consider in some detail the case where the explanatory variable has only two categories. Thus, in this case the data consist of two groups of observations. We may want to study whether there is a difference in income between males and females, or whether there is a difference between Democrats and Republicans on some issue, and so on.

Let us suppose for a while that the country variable only has two categories, Germany and the U.S. In the first group we have five subjective competence scores and in the second group we also have five scores. The question we want to investigate is whether the two groups differ in their values of the dependent variables Y, here subjective competence. Specifically we want to know whether the mean value of the subjective competence scores differs from one country to the next. We could have said that the groups differ if the number of observations differ, or if the standard deviations of Y differ, but here in a presentation of analysis of variance we are concerned with whether the means differ.

One simple way to tell whether the two means differ is to compute the two means and compare the two numbers. Most likely the two numbers are not the same, and that, in a way, answers our question right there.

[12]

But we only have a partial answer because we realize that with another sample of observations from the two countries it may well be that we would have observed another difference between the two means.

But more than that, after looking at the data more closely we may decide that our observed difference between the two means is not large enough to conclude that we really have a difference between the two groups. This is another way of saying that we may not have a statistically significant difference between the two groups. This discussion is illustrated more fully by the three graphs in figure 1.

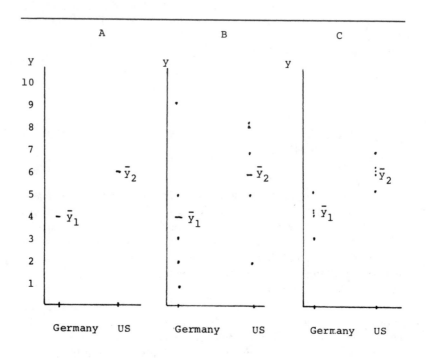

A: Two different group means.

B: Same two means with observations spread far apart. Differences between group means not statistically significant.

C: Same two means with observations close together. Difference between group means statistically significant.

Figure 1: Group Means $\bar{y}_1$ and $\bar{y}_2$ with Single Observations

Graph A shows two numerically different means. From looking at the two means alone we cannot tell whether they are significantly different or not. In graph B the difference between the two means is the same as in graph A, but the difference cannot be trusted very much because the observations themselves are so spread out. Within each of the two groups the observations vary considerably, and because of this we are inclined to say that the difference between the two means is not very convincing.

In graph C the situation is different. The observed difference between the two means is the same as in graphs A and B, but now the observations are clustered around the means in each of the groups. Because the observations are that clustered around their respective means the groups are really separated, and we conclude in graph C that the two means are truly different. Thus, there is a statistically significant difference between the two means in that case. The data in graph C are also displayed in table 1.

The problem for us now becomes how to decide when means are different enough, relative to the spread of the observations in each group, to conclude that there is a statistically significant difference between the means. Analysis of variance helps us answer this question. What we have to do is find a way of measuring numerically how different the means are and how much the observations are spread out around their respective means. With those two measures on hand we are then able to tell whether the means differ significantly or not. But first the difference between two means is analyzed in a more familiar way.

As already noted in chapter 1, there exists a special case of analysis of variance that can be used to investigate the difference between two means. The statistical null hypothesis states that the two population means μ_1 and μ_2 are equal. The test statistic is a t-score where the numerator is

$$(\bar{y}_1 - \bar{y}_2) - (\mu_1 - \mu_2) = \bar{y}_1 - \bar{y}_2 - 0 = 4 - 6 = -2$$

for both cases B and C. Here $\bar{y}_1$ is the sample mean for Germany and $\bar{y}_2$ is the sample mean for the U.S. The denominator is slightly more complicated. First we compute the sample variance of Y in each of the two groups. For the data in case B we get in the first group, with the number of observations equal to n_1,

$$s_1^2 = \Sigma(y_{1j} - \bar{y}_1)^2/(n_1 - 1)$$
$$= \left\{(9{-}4)^2 + (5{-}4)^2 + (3{-}4)^2 + (2{-}4)^2 + (1{-}4)^2\right\}/4$$
$$= 40/4 = 10.00$$

And, similarly, for the observations in the second group, the variance becomes

$$s_2^2 = \Sigma(y_{2j} - \bar{y}_2)^2/(n_2 - 1) = 26/4 = 6.50$$

TABLE 1
Subjective Competence Scores Classified by Country,
Hypothetical Data Case C

	Country	
	Germany	United States
	5	7
	4	5
	4	6
	4	6
	3	6
Mean	4.00	6.00
In symbols:	Germany	United States
	y_{11}	y_{21}
	y_{12}	y_{22}
	y_{13}	y_{23}
	y_{14}	y_{24}
	y_{15}	y_{25}
Mean	$\bar{y}_1$	$\bar{y}_2$

The denominator we seek for the t-statistic can be found from the two variances. Because the numerator is a difference between two means, the denominator becomes

$$s = \sqrt{s_1^2/n_1 + s_2^2/n_2} = \sqrt{10.00/5 + 6.50/5} = 1.82$$

The resulting ratio gives t = −2.00/1.82 = −1.10 with $n_1 + n_2 - 2 = 8$ degrees of freedom. The possible statistical significance of this value can be established from a table of the t-distribution, as discussed below.

With five observations in each group the t-value of −1.10 for the difference between the two means has eight degrees of freedom. Here is the first place where we meet the concept of degrees of freedom. This is a concept that has confused many people. Part of the reason for this confusion is that the formal definition of this concept is heavily mathematical. Another reason is that there exist several equivalent definitions.

Degrees of freedom appear any time we compute a sum of squares. One way to define the concept is to say that the degrees of freedom for a particular sum of squares is equal to the number of terms in the sum we need to know in order to find the remaining terms and thereby complete the sum. One would think that we need all the terms in order to compute a sum. But there are often restrictions imposed on the terms, and these restrictions make it possible to find some of the terms from knowing the remaining ones.

For example, the variance s_1^2 computed above is found from a sum of five squared terms. But there is a restriction on these terms that they add up to zero, because they are all deviations from the mean of the original scores. Therefore, if we were told that the first four terms were 5, 1, −1 and −2, we would know that the fifth term has to be equal to −3 because this is the only number that will make all five terms add up to zero. By squaring each of the five terms and adding the five squares we get a total of 40. Because we initially needed to know only four of the five terms in order to find the sum of 40, we say that this sum has four degrees of freedom. Because this sum is the numerator in the expression for the variance in the first group we also say that the resulting variance has four degrees of freedom. Similarly, the variance in the other group has four degrees of freedom as well. Because we add the two variances to get the denominator for the t-value we also add the corresponding degrees of freedom, and there is therefore a total of eight degrees of freedom for the t-value.

In general, with n_1 observations in one group and n_2 in another there are $n_1 - 1$ degrees of freedom for the variance in the first group and $n_2 - 1$ degrees of freedom for the variance in the second group. When we combine the two variances to get the denominator for the t-expression we add the degrees of freedom and get a total of $n_1 + n_2 - 2$ degrees of freedom for the t-value. We have to realize, however, that when there are n terms in a sum of squares, the degrees of freedom for this sum is not always equal to $n-1$. The sum could have as many as n or as few as one degree of freedom, depending upon the nature of the terms in the sum. The determination of the correct number of degrees of freedom can be difficult. It is simplified, however, by the fact that when two sums of squares add up to a third sum of squares, the degrees of freedom for the two sums also add up to the degrees of freedom for the third sum. This fact is made use of several times in this presentation of analysis of variance.

There is a requirement for this t-test that the variances in the two groups do not differ too much. In other words, s_1^2 should be approximately equal to s_2^2. As an example of what is meant by "approximately equal" we note that, with five observations in each group, we can tolerate having one variance as much as six times as large as the other and still call the variances about equal. With more observations the two variances have to be more alike. For example, when there are 100 observations in each of the two groups, one sample variance cannot be more than 35% larger than the other. For a further discussion of the assumption of homogeneity of variances one can consult books on experimental design and analysis of variance (for example, see Keppel, 1973).

The formal assumption does not state that the two sample variances are equal but that the variances are equal in the two populations from which the two sets of sample observations came. If the assumption of equal population variances is true we should use all the data to estimate this common variance. This is a better use of the data than what we did above where we used one part of the data to find the estimate s_1^2 and another part of the data to find another estimate s_2^2.

One way to find a single estimate is to compute the mean of the two existing estimates, that is

$$s^2 = \tfrac{1}{2}s_1^2 + \tfrac{1}{2}s_2^2$$

One difficulty with this way of combining the two estimates is that when the number of observations in the two groups differ, the variance based on the larger number of observations is a better quantity, in some sense, and should therefore contribute more to the overall s^2. The formula above does not allow for that, because it uses ½ as weights for both variances. It is better to use weights that involve the number of observations in each of the groups, and the overall variance is usually found from the expression

$$s^2 = \frac{n_1 - 1}{n_1 + n_2 - 2} \, s_1^2 + \frac{n_2 - 1}{n_1 + n_2 - 2} \, s_2^2 = \frac{4}{8} \, 10.00 + \frac{4}{8} \, 6.50$$

$$= 8.25$$

When the number of observations in the two groups are equal this formula reduces to the first formula with ½ as weights. When the groups have different number of observations the second formula gives more weight to the variance based on the larger number of observations.

With these changes the denominator for the t-statistic for the difference of two means becomes

$$\sqrt{s^2/n_1 + s^2/n_2} = s\sqrt{1/n_1 + 1/n_2}$$

The t-value itself is found as

$$t = \frac{\bar{y}_1 - \bar{y}_2}{s\sqrt{1/n_1 + 1/n_2}} = \frac{4 - 6}{2.87\sqrt{1/5 + 1/5}} = -1.10 \qquad 8 \text{ df.}$$

which, in this particular case with the same number of observations in the two groups, is the same t-value as presented earlier.

With eight degrees of freedom we need a t-value in excess of plus or minus 2.31 to reject the null hypothesis with a 5% significance level. For the data in case B the hypothesis of equal means therefore cannot be rejected. For the data in case C the means are significantly different because $t = -4.47$. The best estimate of the difference $\mu_1 - \mu_2$ is the sample difference $\bar{y}_1 - \bar{y}_2 = -2$.

Analysis of variance for two groups. The statistical analysis outlined above is the simplest example of an analysis of variance. It is usually not thought of as such, but the formula for t can be changed to become an F-ratio as used in analysis of variance. In order to make the connection between the t-test for the equality of two means and analysis of variance we show below how the formula for t is changed.

First it can be noted that except for not being able to tell whether the sign of t is positive or negative, no other information is lost by squaring the t-value. In the following assume that the two groups have the same numbers of observations, that is, $n_1 = n_2 = n$. Then,

$$t^2 = \frac{(\bar{y}_1 - \bar{y}_2)^2}{s^2(1/n + 1/n)} = \frac{n(\bar{y}_1 - \bar{y}_2)^2}{2s^2} = \frac{5(4-6)^2}{2(8.25)} = 1.21$$

tells us just as much as t itself, except for the sign of the difference $\bar{y}_1 - \bar{y}_2$. But if the null hypothesis being tested is a two-sided hypothesis, the sign does not matter because the null hypothesis of equal population means is tested against the alternative that the two population means are different. In the following we therefore use t^2 instead of t.

Second, it is possible to rewrite the numerator for t^2. It is possible to show that the difference between the two means can be written as the sum of two differences, where each of the new differences is a difference between a group mean and the overall mean $\bar{y}$. Thus,

$$(4-6)^2$$
$$= (\bar{y}_1 - \bar{y}_2)^2 = 2(\bar{y}_1 - \bar{y})^2 + 2(\bar{y}_2 - \bar{y})^2$$
$$= 2(4-5)^2 + 2(6-5)^2$$

Then we can write the expression for t^2 the following way,

$$t^2 = \frac{n(\bar{y}_1 - \bar{y})^2 + n(\bar{y}_2 - \bar{y})^2}{s^2} = \frac{5(4-5)^2 + 5(6-5)^2}{8.25}$$

$$= 1.21$$

In the terminology of analysis of variance this quantity is called F with 1 and $2n - 2$ (=8) degrees of freedom. The numerator measures how much the group means differ from the overall mean. The denominator, which is the common variance in each of the groups, measures how much the observations are spread out around the group means. The same formula also applies when the number of observations in the two groups are not equal. When there are n_1 observations in the first group and n_2 observations in the second group we get

$$F = \frac{n_1(\bar{y}_1 - \bar{y})^2 + n_2(\bar{y}_1 - \bar{y})^2}{s^2} \qquad \begin{array}{l} 1 \text{ and } n_1 + n_2 - 2 \\ \text{degrees of freedom} \end{array}$$

Now it is possible to return to the graphs labeled B and C in figure 1. The numerator for the F-ratio will be the same for the two cases since the two group means $\bar{y}_1$ and $\bar{y}_2$ deviate from the overall mean $\bar{y}$ in the same way. But the denominator s^2 will differ; it will be large for case B and small for case C. In case B the observations are quite far from the means in each of the two groups, and the variance s^2 (=8.25) will therefore be large. This again means that the F-value (=1.21) will be small.

In case C the situation is different. The observations are quite clustered around their group means, and the variance s^2 (=0.50) will therefore be small. The value of F (=20.00) will be large, and we can conclude that a large value of F tells us we have a significant difference between the two group means. Dealing with two groups as we do here, we recall that a t-value larger than 2 or smaller than -2 is usually significant. That means that an F-value larger than about 4 is similarly significant, since the F is obtained by squaring the t. From tables of the F-distribution we find that with many observations and more than two groups the corresponding F-values are significant when they are as small as 2 or 3. In this particular case with two groups and 10 observations, we find from tables of the F-distribution that we need an F-value of 5.32 with a 5% significance level in order to call the observed difference $\bar{y}_1 - \bar{y}_2$ significant.

The value of F is approximately equal to 1.00 when the difference between the means is due to random fluctuations only. The reason for that is as follows. When we have a sample of observations of a random variable with variance s^2 it is known that we can find the variance of the sample mean $(s_{\bar{y}}^2)$ from the equation

$$s_{\bar{y}}^2 = \frac{s^2}{n}$$

Using the data from case B we have $s^2 = 8.25$ and $n = 5$ which gives

$$s_{\bar{y}}^2 = 8.25/5 = 1.65$$

The interpretation of this variance is that if we had a large number of samples from the same population and computed the mean from each sample, then the variance of these sample means would be approximately equal to 1.65.

We are not in the situation here of having many samples, but we do have two means, $\bar{y}_1 = 4$ and $\bar{y}_2 = 6$, and we can find the variance of these two values. Thus, using the usual formula for a variance, another estimate of the variance of the means becomes

$$s_{\bar{y}}^2 = \left\{ (\bar{y}_1 - \bar{y})^2 + (\bar{y}_2 - \bar{y})^2 \right\}/(2-1)$$

$$= \left\{ (4-5)^2 + (6-5)^2 \right\}/1 = 2.00$$

If the differences in the sample means are due to random fluctuations only, then these two estimates of the variance of the means should be about equal. One way to find out how equal the two estimates are consists of taking their ratio, which should be about equal to one. Here

$$F = \frac{2.00}{1.65} = 1.21$$

which is the same F-value as the one above. This ratio is about equal to one, and we conclude that the difference between the means is only due to random fluctuations.

In case C the situation is different. There the F-ratio equals 20.00, and about the only reason it is that large is that the difference between the means is due to something more than random fluctuations. Thus, there is every reason to believe that the two groups of observations are two samples that do not come from the same population. That again implies that we conclude the population means are not equal.

The computation necessary for an analysis of variance as it has been performed above, are usually summarized in a table like table 2 which shows that data from case B. The numbers on the first line in the table refer to the group means. The difference between the group means is measured by the sum of squares.

$$n_1(\bar{y}_1 - \bar{y})^2 + n_2(\bar{y}_2 - \bar{y})^2 = 5(4-5)^2 + 5(6-5)^2 = 10.00$$

This sum is sometimes called the between group sum of squares.

Because we here have two groups we divided above by $2-1 = 1$ in order to get the variance of the means. Another way of saying the same thing is that the between group sum of squares is based on one degree of freedom. In the general case the number of degrees of freedom for the between group sum of squares is based on one degree of freedom. In the general case the number of degrees of freedom for the between group sum of squares equals one less than the number of groups. By dividing a sum of squares by its degrees of freedom we get what is called the corresponding mean square. Here the between group mean square becomes $10.00/1 = 10.00$.

The next to the last entry on the first line is the F-ratio measuring whether there is a systematic difference between the group means or not.

TABLE 2
Analysis of Variance Table for the Data in Case B

Source	Sum of squares	Degrees of freedom	Mean square	F-ratio	Significance
Between groups	10.00	1	10.00	1.21	0.30
Within groups	66.00	8	8.25		
Total	76.00	9			

The last entry on the first line gives the probability of observing an F-value as large as the one we got or larger under the assumption that the population means are equal. When this probability is less than 0.05, or 0.01 or whatever significance level we choose, we conclude that the observed difference between the sample means is significant and that the population means are therefore different. One has to realize, however, that a small probability only helps us establish that the population means are different, it says nothing about how different they may be.

The numbers on the second line refer to the variation in the observations within the two groups, leading to the common estimate s^2 of the variance in the two groups. We have seen above that the numerator for s^2 measures how much the observations differ from their group means. When we add up the deviations of the observations from their means we get

$$(\bar{y}_{11} - \bar{y}_1)^2 + (\bar{y}_{12} - \bar{y}_1)^2 + \ldots + (\bar{y}_{15} - \bar{y}_1)^2$$
$$+ (\bar{y}_{21} - \bar{y}_2)^2 + (\bar{y}_{22} - \bar{y}_2)^2 + \ldots + (\bar{y}_{25} - \bar{y}_2)^2$$
$$= (9 - 4)^2 + (5 - 4)^2 + (3 - 4)^2 + (2 - 4)^2 + (1 - 4)^2$$
$$+ (8 - 6)^2 + (8 - 6)^2 + (7 - 6)^2 + (5 - 6)^2 + (2 - 6)^2 = 66.00$$

Since the sum of squared deviations in each group is based on $n - 1 = 4$ degrees of freedom, the sum of squares above has 8 degrees of freedom. In the general case the number of degrees of freedom for the sum of squares measuring the variations within the groups equals the total number of observations minus the number of groups. Finally, the common variance of the dependent variable Y in each of the two groups is obtained by dividing the within group sum of squares by its degrees of freedom, thereby giving us the within group mean square of $66.00/8 = 8.25$. The F-value is then obtained by dividing the two mean squares.

The last line in table 1 does not have much use in the analysis, even though it is sometimes easier to find the within group sum of squares as the difference between the total sum of squares and the between group sum of squares rather than directly, as we did above. The total sum of squares is found by subtracting the overall mean from each of the observations, squaring all these differences and then adding them all up. The overall mean here is 5, and the total variation of the observations around this mean becomes

$$
\begin{aligned}
&(y_{11} - \bar{y})^2 + (y_{12} - \bar{y})^2 + \ldots + (y_{15} - \bar{y})^2 \\
&+ (y_{21} - \bar{y})^2 + (y_{22} - \bar{y})^2 + \ldots + (y_{25} - \bar{y})^2 \\
&= (9 - 5)^2 + (5 - 5)^2 + (3 - 5)^2 + (2 - 5)^2 + (1 - 5)^2 \\
&+ (8 - 5)^2 + (8 - 5)^2 + (7 - 5)^2 + (5 - 5)^2 + 2 - 5)^2 = 76.00
\end{aligned}
$$

With 10 observations this sum has nine degrees of freedom, and in the general case the number of degrees of freedom for the total sum of squares equals one less than the total number of observations. This total sum of squares can also be computed using the common computational formulas for a variance where one does not have to subtract the mean from each observation.

Statistical theory. There is a formal mathematical theory that underlies the analysis we have done above, and while we do not present the full theory here it is important to realize some aspects of that theory. Part of the theory is presented below and another part is presented at the end of this chapter.

The analysis is founded on the assumption that we can decompose each observation into three additive terms; that is, we have to be able to write each observation as a sum of three terms. The decomposition can be written

observation = overall mean

+ deviation of group mean from overall mean

+ deviation of observation from group mean

The overall mean is a constant, common to all the observations. The deviation of a group mean from the overall mean is taken to represent the effect on each observation from belonging to that particular group. In our example the effect of belonging to the first group is $\bar{y}_1 - \bar{y} = 4 - 5 = -1$, and the effect of belonging to the second group is measured as $\bar{y}_2 - \bar{y} = 6 - 5 = 1$. The deviation of an observation from its group mean is taken to represent the effect on that observation of all variables other than the group variable, here country. For the data in case C these effects on the various observations are

$$-1, 0, 0, 0, 1 \quad \text{and} \quad -1, 0, 0, 0, 1$$

These terms are also called the residuals. Not surprisingly, the sum of the residuals equals zero in each group. The sum of the squared residuals equals 4, and this is the sum we also have called the within group sum of squares.

For the 10 observations in case C we now get the decompositions

$$
\begin{array}{lll}
3 = 5 - 1 - 1 & \text{and} & 5 = 5 + 1 - 1 \\
4 = 5 - 1 + 0 & & 6 = 5 + 1 + 0 \\
4 = 5 - 1 + 0 & & 6 = 5 + 1 + 0 \\
4 = 5 - 1 + 0 & & 6 = 5 + 1 + 0 \\
5 = 5 - 1 + 1 & & 7 = 5 + 1 + 1
\end{array}
$$

The general expression becomes

$$\bar{y}_{ij} = \bar{y} + (\bar{y}_i - \bar{y}) + (y_{ij} - \bar{y}_i)$$

where i equals 1 or 2 and refers to the two groups, and j equals 1, 2, 3, 4 or 5 and refers to the observation within each group.

In order to perform the F-test the way we have done, the residuals have to satisfy the assumption that their distribution follows a normal distribution. The histogram of the 10 residuals for the data in case C is shown in figure 2. While we do not have a normal distribution in that figure, we do have a unimodal, bellshaped histogram satisfactorily indicating that we may well have an underlying normal distribution of residuals.

The residuals will not always display a distribution that follows a normal distribution as closely as the distribution shown in figure 2. With only moderate departures from normality experience has shown that the statistical tests discussed here are not affected. Another way of saying the

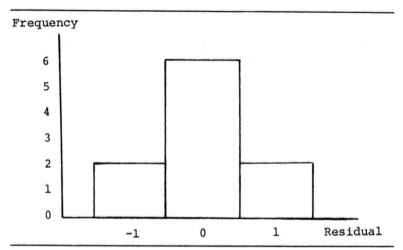

Figure 2: Histogram of the Residuals for the Data in Case C

same thing is to say that t-tests and F-tests are robust tests. When the departure from normality is more severe it may be possible to change the original observations of the dependent variable and thereby make the distribution of the residuals more like a normal distribution.

Such transformations of the observations can be used to make heavily skewed distributions more symmetric. For example, a skewed distribution with many small and fewer large observations will become more symmetric if we use the square root of each observation rather than the original observations themselves. Other types of distributions may be more problematic. A u-shaped distribution, for example, cannot be changed into a bell-shaped distribution by any kind of transformation. In such cases one must be much more careful in one's interpretation of the significance level used in the tests. The distribution of the residuals is discussed further in chapter 3 in the section on the residual variable.

However, there is a complication with what we have done here. The differences $\bar{y}_1 - \bar{y}$ and $\bar{y}_2 - \bar{y}$ are only estimates of the true effects on the dependent variable from the observations belonging to the two groups. Similarly, the residuals we have found are only estimates of the effects of all other variables on the dependent variable. Finally, $\bar{y}$ is only an estimate of the overall level of the dependent variable.

The formal model, which specifies an additive model involving the true effects, can be written

$$y_{ij} = \mu + \alpha_i + \epsilon_{ij}$$

As before, i refers to the group and j refers to the observation within the group. In this model μ is a constant and refers to the overall level of the dependent variable, α_1 and α_2 represent the effect on an observation of belonging to group 1 and 2, respectively. ϵ_{ij} is the effect on the j-th observation in the i-th group of all other variables, and in our example we have 10 such terms ϵ_{11}, ϵ_{12}, ϵ_{13}, ϵ_{14}, ϵ_{15}, ϵ_{21}, ϵ_{22}, ϵ_{23}, ϵ_{24} and ϵ_{25}. These 10 residual terms are assumed to come from a normal distribution with mean zero and variance σ^2.

Instead of having μ, the two α's and the ten ϵ's we have ten observations (the y's) located in two groups. These observations are used to find estimates of the true, unknown effects designated by Greek letters. Thus, we can never verify whether the true effects are additive and whether the true residuals have a normal distribution. But we can look at the distribution of the estimated residuals and hope that if this distribution looks fairly normal, then the distribution of the true residuals is not too far from normal. One should always find the estimated residuals and examine their distribution for normality. The distribution of the residuals is discussed further in capter 3 in the section on the residual variable.

More Than Two Groups

Analysis of variance. We can now turn to the case where the independent variable, here country, has more than two categories. Other examples, if the independent variable is political party we may have Democrat, Republican, independent, other parties, and apolitical as the categories. With religion we may have Protestant, Catholic, Jew, other and none. Each category defines a group of observations of the dependent variable, and rather than two groups we now have k groups in the general case.

As before, we want to investigate whether the groups differ with respect to the dependent variable Y. In the Almond and Verba study there are observations from five countries; Great Britain, Italy, Mexico, the United States, and West Germany. We want to know whether the five countries differ in their level of subjective competence. From the samples[1] taken within each of the five countries we have the sample means of subjective competence as shown in table 3.[2]

It is still not enough to simply compute the group means and examine whether they are different or not. While the means differ in numerical values we still have to investigate whether the differences are simply random variations that occurred by chance, or whether there are systematic differences between the means. That is, we have to compare the variation in the means with how much the observations vary within each of the

TABLE 3
Mean Subjective Competence Scores and Number of
Observations from Five Countries

Country	Subjective competence Mean	Number of observations
Great britain	0.17	918
Italy	-0.33	762
Mexico	0.05	828
US	0.36	940
West Germany	-0.21	867
Total	0.02	4315

groups. This is the very same type of reasoning we went through in the discussion with two groups.

For a numerical example we turn to the case with five groups. For the sake of simplicity and without any loss of generality we restrict ourselves to a small number of cases in each group. The numerical values are all hypothetical and constructed to be easy to work with. The data for this section are found in figure 3, showing five groups and a total of 18 observations. The groups no longer have the same number of observations, but with only one independent variable it makes little difference for the analysis whether the number of observations in the groups are the same or not.

The analysis of these data proceeds as follows. The 18 observations have different values, and one way to measure how different they are is to subtract the overall mean $\bar{y} = 5$ from each observation, square all the differences and add them up. This gives the total sum of squares, TSS, where

$$
\begin{aligned}
\text{TSS} &= \Sigma(y_{ij} - \bar{y})^2 \\
&= (3-5)^2 + (4-5)^2 + (4-5)^2 + (4-5)^2 + (5-5)^2 \\
&\quad + (5-5)^2 + (6-5)^2 + (6-5)^2 + (6-5)^2 + (7-5)^2 \\
&\quad + (1-5)^2 + (3-5)^2 \\
&\quad + (6-5)^2 + (7-5)^2 + (8-5)^2 \\
&\quad + (4-5)^2 + (5-5)^2 + (6-5)^2 = 50
\end{aligned}
$$

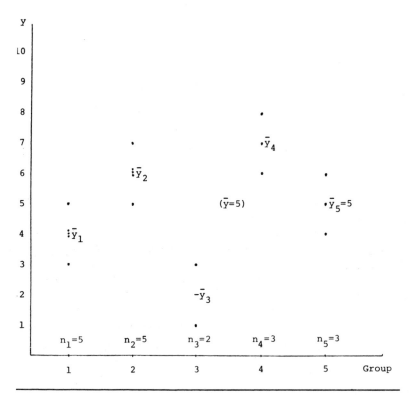

Figure 3: Five Groups of Observations with Means and the Number of Observations

One should note that there is a simpler way to compute this sum. An equivalent formula is

$$TSS = \Sigma y_{ij}^2 - (\Sigma y_{ij})^2/n$$
$$= (3^2 + 4^2 + \ldots + 6^2) - (3 + 4 + \ldots + 6)^2/18$$
$$= 500 - 90^2/18 = 50.00$$

With this formula one does not have to subtract the overall mean from each observation before squaring.

There are 18 terms in the first sum for the total sum of squares above. The first number is $3 - 5 = -2$ before squaring, the second is $4 - 5 = -1$, and so on. Because there are 18 such numbers here the total sum of squares has 17 degrees of freedom. By that is meant that we need only know 17 of the numbers in order to find all 18 of them. If we know that the first 17 numbers are

$$-2 \ -1 \ -1 \ -1 \ 0 \ 0 \ 1 \ 1 \ 2 \ -4 \ -2 \ 1 \ 2 \ 3 \ -1 \ 0$$

then we know from these numbers that the 18th number has to equal 1. That is because the sum of all 18 numbers equals zero, and the sum of the 17 numbers above equals -1. Thus, the last number has to equal 1. The total sum has to equal zero because we are dealing with deviations from the overall mean.

Because the last number was "hidden" in the other 17 numbers, that number does not tell us anything new that we did not already know. Therefore, even though the total sum of squares TSS is computed from 18 numbers, it is only based on 17 degrees of freedom. In the general case the number of degrees of freedom for the total sum of squares equals the total number of observations minus one.

The total variation of the scores of 50.00 has two sources. One is the variation of the group means from the overall mean. The magnitude of this variation is measured by the between group sum of squares, BSS. Here we have

$$BSS = \Sigma n_i (\bar{y}_i - \bar{y})^2$$

$$= 5(4-5)^2 + 5(6-5)^2 + 2(2-5)^2 + 3(7-5)^2 + 3(5-5)^2$$

$$= 40.00$$

This sum is based on four degrees of freedom because there are five groups altogether. In the general case the number of degrees of freedom equals one less than the number of groups.

We note that this sum of five terms is a direct generalization of the case when we had only two groups. In that case the between group sum of squares contained two terms. If the country variable made no difference for subjective competence, the mean subjective competence scores would be about the same in each of the countries. But here the means are not the same, and one way to measure how different they are is to compute the between group sum of squares the way it has been computed here. The magnitude of that sum tells us how large the effect of the country variable is on the subjective competence variable.

The other source of variations in the scores is the variation of the scores from their respective group means. The difference between an observation and the mean of the group to which it belongs has been called the residual.

For our example the sum of the squared residuals, RSS, should equal 50.00 − 40.00 = 10.00 on 17 − 4 = 13 degrees of freedom. The 18 residuals in our example are shown in table 4. The table also shows that the sum of squared resiudals equals 10.00, as expected. The residual sum of squares here is also a direct generalization of the residual sum of squares we had with only two groups.

The residual sum of squares measures the effect on the dependent variable of all variables other than the group variable. If these variables had no effect on the subjective competence scores, then all the observations in a given country would have the same value. But the scores in each of the countries are not alike, and one way to measure how different they are is to subtract the group mean from each observation, square these differences, and add up the squares. The magnitude of this sum, the residual sum of squares, tells us how large the effect is of all these variables on the dependent variable. The residual degrees of freedom always equals the number of observations minus the number of groups.

The results obtained above can be displayed as before in an analysis of variance table, as shown in table 5. In that table the BSS and RSS are divided by their degrees of freedom to get the corresponding mean squares. The mean square for the group variable equals 10.00, and the mean square for the residual variable equals 0.77. That means that the variance of the 18 residuals equals 0.77, and the standard deviation of the residuals equals 0.88 One way to interpret that standard deviation is that the average deviation of the observations from the means of their groups equals 0.88.

TABLE 4
Residuals Arranged by Group, Their Sum of Squares and Degrees of Freedom

Group	Residuals	Sum of squares	Degrees of freedom
1	−1 0 0 0 1	2.00	4
2	−1 0 0 0 1	2.00	4
3	−1 1	2.00	1
4	−1 0 1	2.00	2
5	−1 0 1	2.00	2
	Total	10.00	13

TABLE 5
Analysis of Variance Table for the Data in Figure 3

Source	Sum of squares	Degrees of freedom	Mean square	F-ratio	Signi- ficance
Groups	40.00	4	10.00	13.00	0.0002
Residual	10.00	13	0.77		
Total	50.00	17			

$E^2 = 0.80$

The F-ratio is again obtained by dividing the mean square for groups by the mean square for the residual variable. As before, a value around 1.00 tells us that the variation in the group means is only a random varia- tion from sample to sample. When the F-ratio is a good deal larger than 1.00 we can conclude that the variation in the group means is more than what could be expected by chance, and the means in the populations from which these samples came are therefore different. Here we have F = 13.00, and from a table of the F-distribution we find that with 4 and 13 degrees of freedom we need a value of F larger than 5.20 in order to conclude that the sample means are significantly different, using a 5% significance level. If the five population means really are the same, the probability of getting an F-value of 13.00 or larger in a sample is only 0.002. This probability is so small that our conclusion becomes that the population means are different. The immediate question is how different they are, and to get some idea of the magnitudes of the differences we can take the sample means as estimates of the population means. That way the differences between the sample means become estimates of the differences between the population means.

As before, we can examine the residuals for any unusual pattern and to what extent they follow a normal distribution. A histogram of the 18 re- siduals is very much like the one in figure 3, and it shows how we have a variable with a mean of zero and a symmetrical, unimodal distribution. The distribution of the residuals is discussed further in chapter 3 in the section on the residual variable.

The correlation ratio. The effect of the country variable on subjective competence can be measured in two different ways. The effect of the vari- able in a particular country can be measured as the mean subjective com- petence score in that country minus the overall mean. The total effect

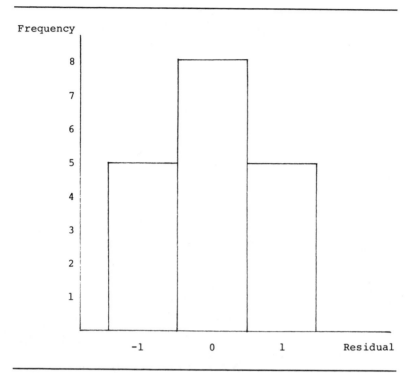

Figure 4: Histogram of 18 Residuals for the Data in Figure 3

of the country variable can be measured as the group (between) sum of squares. But we do not yet have any measure of the relationship between the explanatory and the dependent variable that resembles a correlation coefficient, giving the strength of the relationship.

One way to get such a measure is as follows. If we had all the values of Y and were asked to estimate the value of a particular respondent without being told what the score is for that respondent, one possible estimate would be the overall mean $\bar{y}$. This is a value that lies in the middle of all the observations, and in the absence of any information about the respondent it seems reasonable to choose such a central value as our estimate. If the true score is y_{ij}, then the error we made becomes $y_{ij} - \bar{y}$. Suppose the "penalty" assessed for this error is the square of the error; that is $(y_{ij} - \bar{y})^2$. If we did this for all the respondents and used the mean as an estimate each time, then the total "penalty" would be the sum of all squared errors, $\Sigma(y_{ij} - \bar{y})^2$. This sum is the total sum of squares in analysis of variance. From table 5 we have TSS = 50.00 as a measure of our ability to predict the observations in our numerical example.

Next we change the guessing game, and now we are told what group an observation belongs in. Our predicted value would then be the mean of the observations in that group, and for the i-th group this mean is denoted $\bar{y}_i$. The error we make now is $y_{ij} - \bar{y}_i$. The "penalty" for making this error still is the square of the error, $(y_{ij} - \bar{y}_i)^2$. The total "penalty" doing this for all the observations becomes the sum $\Sigma(y_{ij} - \bar{y}_i)^2$. This is the residual sum of squares in analysis of variance. In our example RSS = 10.00 as a measure of how well we can predict when we know the group the observation belongs to.

We can see that knowing the group improves our prediction. The penalty is reduced when we know to which group each observation belongs. For these data the improvement equals $50.00 - 10.00 = 40.00$. The relative improvement becomes

$$E^2 = \frac{50.00 - 10.00}{50.00} = \frac{TSS - RSS}{TSS} = 0.80$$

This quantity is called the correlation ratio and is sometimes denoted as eta squared. It is the ratio of the sum of squares for the explanatory variable to the total sum of squares. It tells us how much of the variation in the dependent variable is explained by the explanatory variable, in the sense of how much our prediction is improved by knowing the group when the "penalty" is measured by the square of the prediction error.

The same quantity is computed in regression analysis and is there called R^2, where R is the multiple correlation coefficient. When there is only one explanatory variable in the regression analysis, the multiple correlation coefficient reduces to the ordinary correlation coefficient relating the explanatory and the dependent variable.

Formal model. It is now possible to return to the formal model for a fixed effect one-way analysis of variance in somewhat greater detail. The model specifies that an observed value of the dependent variable can be written as a sum of three components. That is,

Observed value of Y = constant

+ effect of being in a particular group (effect of the independent variable X)

+ effect of all other variables (residual)

For the j-th observation in the i-th group the model can be translated into symbols and becomes

$$y_{ij} = \mu + \alpha_i + \epsilon_{ij}$$

where μ is the constant, α_i is the effect of being in group i and ϵ_{ij} is the effect on that observation of all other variables. As usual with a statistical model of this kind we want to find estimates of the parameters and investigate the effect of the independent variable on the dependent variable.

For the 18 observations in our example the model becomes

$$y_{11} = \mu + \alpha_1 \qquad\qquad + \epsilon_{11}$$
$$\begin{array}{ccc} \cdot & \cdot & \cdot \\ \cdot & \cdot & \cdot \\ \cdot & \cdot & \cdot \end{array} \qquad\qquad \begin{array}{c} \cdot \\ \cdot \\ \cdot \end{array} \quad \text{Group 1}$$
$$y_{15} = \mu + \alpha_1 \qquad\qquad + \epsilon_{15}$$

$$y_{21} = \mu + \alpha_2 \qquad\qquad + \epsilon_{21}$$
$$\begin{array}{ccc} \cdot & \cdot & \cdot \\ \cdot & \cdot & \cdot \\ \cdot & \cdot & \cdot \end{array} \qquad\qquad \begin{array}{c} \cdot \\ \cdot \\ \cdot \end{array} \quad \text{Group 2}$$
$$y_{25} = \mu + \alpha_2 \qquad\qquad + \epsilon_{25}$$

$$y_{31} = \mu \qquad + \alpha_3 \qquad + \epsilon_{31} \quad \text{Group 3}$$
$$y_{32} = \mu \qquad + \alpha_3 \qquad + \epsilon_{32}$$

$$y_{41} = \mu \qquad\quad + \alpha_4 \quad + \epsilon_{41}$$
$$y_{42} = \mu \qquad\quad + \alpha_4 \quad + \epsilon_{42} \quad \text{Group 4}$$
$$y_{43} = \mu \qquad\quad + \alpha_4 \quad + \epsilon_{43}$$

$$y_{51} = \mu \qquad\qquad + \alpha_5 + \epsilon_{51}$$
$$y_{52} = \mu \qquad\qquad + \alpha_5 + \epsilon_{52} \quad \text{Group 5}$$
$$y_{53} = \mu \qquad\qquad + \alpha_5 + \epsilon_{53}$$

There is a restriction on the α's in this model, so that when each of them is multiplied by the number of observations in the corresponding group then the sum of these products equals zero. In our example,

$$5\alpha_1 + 5\alpha_2 + 2\alpha_3 + 3\alpha_4 + 3\alpha_5 = 0$$

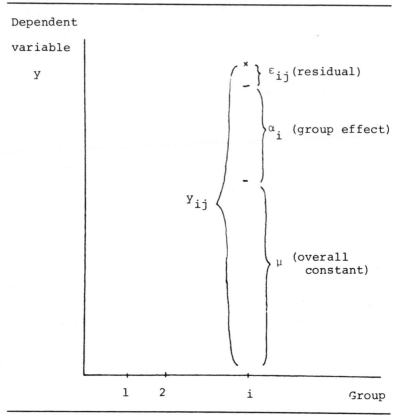

Figure 5: Decomposition of the j-th Observation in the i-th Group into Three Components

Figure 5 shows how a typical observation is decomposed into the three components of this model.

The first task is to find numerical estimates of the parameters in the model. Above we have 18 equations, and with one μ, 5 α's and 18 ϵ's there are 24 unknowns. Because of the restriction above on the α's the last one can be found if we know the other four. But we still have 23 unknowns and only 18 equations, and we cannot solve the equations for the unknowns directly.

Instead, we used the method of least squares to estimate the unknown parameters from the known data. The method uses as estimates the values of the parameters that make the sum

$$\Sigma(y_{ij} - \mu - \alpha_i)^2$$

as small as possible. The estimates are denoted by Greek letters with a "ˆ" above. One can show that the smallest possible sum is obtained when

$\hat{\mu} = \bar{y}$ overall mean

$\hat{\alpha}_i = \bar{y}_i - \bar{y}$ deviation of group mean from overall mean

$\hat{e}_{ij} = y_{ij} - \bar{y}_i$ deviation of observation from group mean

For our example we get the numerical values

$$\hat{\mu} = \bar{y} = 5 \qquad \hat{\alpha}_1 = \bar{y}_1 - \bar{y} = 4 - 5 = -1$$
$$\hat{\alpha}_2 = \bar{y}_2 - \bar{y} = 6 - 5 = 1$$
$$\hat{\alpha}_3 = \bar{y}_3 - \bar{y} = 2 - 5 = -3$$
$$\hat{\alpha}_4 = \bar{y}_4 - \bar{y} = 7 - 5 = 2$$
$$\hat{\alpha}_5 = \bar{y}_5 - \bar{y} = 5 - 5 = 0$$

The 18 estimated residuals have already been listed above. With these estimated parameters the sum of squares above is equal to 10.00, and there is no way one can have values of the parameters that will make that sum any smaller.

When the residuals have a normal distribution we can compute the F-ratio and use it to test the hypothesis that the α's are all equal to zero. That is the statistical hypothesis which states that the population means in the five countries are all equal, and that country as a variable has no effect on subjective competence. The formal reason we compute the F-ratio is now as follows. Let the variance of the true residuals be denoted σ^2. In order to estimate this variance we compute the variance of the estimated residuals. We get $s^2 = 10.00/13 = 0.77$, where 10.00 is the residual sum of squares and 13 is the corresponding number of degrees of freedom. This is the denominator for F.

The numerator is more complicated. It has been computed above as the sum

$$n_1(\bar{y}_1 - \bar{y})^2 + n_2(\bar{y}_2 - \bar{y})^2 + n_3(\bar{y}_3 - \bar{y})^3 + n_4(\bar{y}_4 - \bar{y}) + n_5(\bar{y}_5 - \bar{y})^2$$
$$= n_1\hat{\alpha}_1^2 + n_2\hat{\alpha}_2^2 + n_3\hat{\alpha}_3^2 + n_4\hat{\alpha}_4^2 + n_5\hat{\alpha}_5^2 = 40.00$$

It can be shown that this number is an estimate of the quantity

$$(5-1)\sigma^2 + (n_1\alpha_1^2 + n_2\alpha_2^2 + n_3\alpha_3^2 + n_3\alpha_3^2 + n_4\alpha_4^2 + n_5\alpha_5^2)$$

where 5 here is the number of groups we are dealing with. What we have here is the between group sum of squares. By dividing by the number of degrees of freedom, in this case 4, we get the between group mean square. It becomes an estimate of the quantity

$$\sigma^2 + (n_1\alpha_1^2 + n_2\alpha_2^2 + n_3\alpha_3^2 + n_4\alpha_4^2 + n_5\alpha_5^2)/4$$

The numerical estimate of this expression in our example is $40.00/4 = 10.00$. Finally, but taking the ratio

$$F = \frac{\text{Between Mean Square}}{\text{Within Mean Square}} = \frac{10.00}{0.77} = 13.00$$

we have an estimate of the ratio

$$\frac{\sigma^2 + (5\alpha_1^2 + 5\alpha_2^2 + 2\alpha_3^2 + 3\alpha_4^2 + 3\alpha_5^2)/4}{\sigma^2}$$

$$= 1.00 + (5\alpha_1^2 + 5\alpha_2^2 + 2\alpha_3^2 + 3\alpha_4^2 + 3\alpha_5^2)/4\sigma^2$$

There are now two possibilities; that all the α's really equal zero or that not all the α's equal zero. If all the α's really equal zero, then the last expression above equals 1.00. The estimate we have obtained of 13.00 is then a very bad estimate of 1.00. On the other hand, if two or more of the α's do not equal zero, then the value of the theoretical expression above is something more than 1.00. It may, in that case, be that 13.00 is not such a bad estimate of that expression after all.

Thus, with F-values larger than the cutoff values given in the F-table, we conclude that at least some of the effects of the nominal variable, as measured by the α's, are different from zero. The formal statistical null hypothesis states

$$H_0: \alpha_1 = \alpha_2 = \ldots = \alpha_k \ (=0)$$

and it is rejected for large values of F. But even a large value of F does not tell us how large the effects are, and after the hypothesis is rejected one needs to estimate the effects in order to see how large they are. Obviously, small effects, even though they may be statistically significant, are not as interesting substantively as large effects.

If the α's are equal to zero the hypothesis is true, and the groups are all alike. In that case the observed F-value should be near 1.00. There will be variations in the F-value from sample to sample, and the range of possible values are given by the tables of the F-distribution.

Computing and numerical results. Unless one has a small number of observations the computations necessary for a one-way analysis of variance are usually performed on a computer. All standard statistical program packages have a program for such an analysis. In such program packages the necessary programming has already been done. All the user has to do is to specify on a few punch cards how the data are organized and how much of the possible output is wanted. One commonly used program package in the social sciences is the SPSS (1975) package, and it contains a program called ONEWAY. In order to use this program one needs the data plus two additional cards. One card is for the independent variable, and it contains the number of categories and the variable name. The other card contains the name of the dependent variable. The output usually contains the mean, standard deviation, range and number of observations for each group, and the whole dataset plus the various sums of squares, mean squares, and F-ratio. Sometimes one also gets the significance of the F-ratio. Our computations were made using programs from the OSIRIS (1973) package of programs.

The OSIRIS program F-means performs one-way analysis of variance, and the output is presented on a single page. For each category (group, treatment) of the explanatory variable, which here is called control variable, the output gives the mean and standard deviation of the dependent variable. Below this display one can find the relevant items of analysis of variance such as the total sum of squares, within sum of squares, between sum of squares, the F-ratio with degrees of freedom, and the correlation ratio. Although these items are not arranged in the conventional analysis of variance table, such a table can still easily be constructed from the given results. In particular, the mean squares are not given, but they can be found by dividing the sums of squares by their degrees of freedom.

One should note that numbers on computer outputs are often given in the so-called E-notation, for example something like 0.3663934E 04. To get the answer we are here asked to multiply the number in front of the E by the exponent 4. In other words, we must move the decimal point four places to the right. Doing that we get the number above equal to 3663.934.

When the E is given with a minus, for example E-03, the decimal point should be moved to the left, here three places.

Some of these results for our example on subjective competence and five countries are shown in table 6. The five means are given in table 3, and the F-ratio of 84.24 shows that the means are significantly different. If the five population means are really equal, the probability of getting an F-value of this size or larger is very, very small, and we therefore conclude that the population means are not equal. Not equal can then mean that they are only a little bit different or that they are very different. The largest difference between two countries is between Italy ($\bar{y} = -0.33$) and the United States ($\bar{y} = 0.36$).

To get some feeling for the magnitude of this difference we would like to know about the range of values of the observations in the two countries. The standard deviation of the observations is 0.89 in each of the two countries, as shown below. For almost any set of observations most of the observations lie within two standard deviations from the mean. The range of values in Italy is therefore approximately from -2.11 to 1.45, while the range of values in the United States is approximately from -1.42 to 2.14. Thus, there is a considerable overlap of scores in the two countries, but the mean in the United States is about 0.69 larger than the mean in Italy. Similar comparisons can be made between other countries.

The standard deviation of the observations in each of the countries is found as the square root of the residual mean square. The value here is

TABLE 6
Analysis of Variance Table for Subjective Competence Scores from Five Countries

Source	Sum of squares	Degrees of freedom	Mean squares	F- ratio	Signi- ficance
Countries	265.69	4	66.423	84.24	0.00
Residual	3398.24	4310	0.789		
Total	3663.93	4314			

$E^2 = 0.07$

$\hat{\alpha}_{GB} = 0.170 - 0.023 = 0.147$

$\hat{\alpha}_{It} = -0.327 - 0.023 = -0.350$

$\hat{\alpha}_{Mx} = 0.047 - 0.023 = 0.024$

$\hat{\alpha}_{US} = 0.357 - 0.023 = 0.334$

$\hat{\alpha}_{WG} = -0.209 - 0.023 = -0.232$

0.79, and this is the variance of the residuals. The standard deviation becomes 0.89. Since a residual is the difference between an observation and the group mean, this standard deviation measures the spread of the observations in each group. Under the assumption of equal variances in the groups 0.89 can be used as the standard deviation for each of the groups.

With a large number of observations and a correspondingly large number of degrees of freedom for the residual sum of squares, the F-value in an analysis of variance is often quite large, even if the means may not differ very much. One should therefore always examine the magnitudes of the differences between the sample means to see whether they are substantively, as well as statistically, significant. In addition, the correlation ratio tells something about the relationship between the two variables. Here this ratio is as small as 0.07, and we conclude that the relationship between country and subjective competence is fairly weak.

3. TWO-WAY ANALYSIS OF VARIANCE, ALL CATEGORIES

Unrelated Explanatory Variables

Introducing one more variable. The five nations examined by Almond and Verba were found in the last chapter to differ significantly with regard to level of subjective competence. This may suggest that unique factors of history and culture are responsible for these national differences. On the other hand, one could argue that such differences reflect differences in economic and social development, with rural-agrarian nations exhibiting a generally lower level of subjective competence than urban-industrial ones. As a nation moves from the former type to the latter type of society, its overall level of subjective competence rises.[3]

We are now dealing with two hypotheses: one asserts that national experiences account for differences in subjective competence, and the other asserts that such differences are explained by the level of development. In order to investigate the effects of these two variables on subjective competence we need to have some measure of development for each respondent in the study. In the following we let education be such a measure.

The respondents are now classified according to two variables, country and education. Such a classification allows us to compare nations as well as educational groups with respect to level of subjective competence. Furthermore, mean scores on subjective competence within a given educational group can be compared across nations, and means within a given country can be compared across educational groups. With two or more explanatory variables in an experiment one has what is called a factorial design.

Table 7 displays these means by country and education. By reading across the rows of this table one can sense the differences in subjective competence which exist between the various levels of education within a given country. Surely these differences emerge with similar strength and regularity in each of the rows (countries). On the other hand, by focusing on the columns of the table one can get an impression of the differences in subjective competence which emerge between national groups matched in education. The hypothesis of no national differences when development is held constant would require national groups on any given level of development (education) to be identical in subjective competence.

The means displayed in table 7 are not identical. But such an observation does not resolve the issue. What we have to know is whether the observed national differences within the "educational columns" are more than random variations and whether they are as strong as those which were observed in the one-way analysis. These questions cannot be answered by glances at the cell means. The number of observations in each category is given in table 16.

We are using this example to illustrate two-way analysis of variance. Strictly speaking, a two-way analysis of variance calls for the two explana-

TABLE 7
Subjective Competence Means by Country and Education

| | EDUCATION | | | |
COUNTRY	Grade School	High School	College	Mean
Great Britain	0.06	0.33	0.41	0.16
Italy	-0.40	-0.22	0.45	.-0.32
Mexico	-0.01	0.34	0.40	0.05
United States	-0.10	0.48	0.87	0.36
West Germany	-0.29	0.17	0.45	-0.20
Mean	-0.16	0.33	0.69	0.02

tory variables to be nominal level variables, but education is at least an ordinal level variable. Neglecting the given order of the categories of the education variable by using analysis of variance, one loses the ability to detect trends in the data across the categories. But analysis of variance can still be used as a first and somewhat rough tool for the investigation of patterns in the data.

Two examples are used in this chapter. The results of a two-way analysis of variance of the Almond and Verba data are given for five countries and three educational levels. In order to explain the methods more fully we also use a small example with hypothetical data. In order to keep this numerical example small we restrict the presentation to two countries and two levels of education, with a total of 10 observations.

Suppose we have the data in table 8. These are the same data as in case C of the previous chapter, with the difference that the observations are now also classified according to education. Basically, we want to know whether these 10 scores have been allocated to the four cells in the table completely by chance or whether country and education had any effect on the dependent variable. The table also shows the numbers replaced by symbols. Three subscripts are needed in order to show where an observation is located. The first subscript refers to the row, the second to the column and the third counts the observations within a given cell. Since all the observations in the first row have 1 as the first subscript, the mean for those observations has 1 as the first subscript; and similarly for the second row and the two columns.

Types of effects. We have already found that the two group means of 4.00 and 6.00 for Germany and the United States, respectively, differ in a significant way and by more than what can be expected by chance alone. From that we conclude that the country variable has an effect on the dependent variable. From the two education means it also looks like a high education produces a higher subjective competence score than a low education, but we do not know yet whether the difference in the two means 5.25 and 4.83 could have occurred by chance or not. Below we examine whether the education variable does have an effect on subjective competence.

In addition to the country and education effects there is a possible third effect on the subjective competence scores. The number in parenthesis in each cell in table 8 is the mean of the observations in that cell, and these means show here the existence of an effect on the subjective competence scores over and beyond the separate effects of country and education. We are referring to the fact that in Germany a high education

TABLE 8
Subjective Competence Scores Classified by Country and Education, Hypotetical Data

Country

		Germany	United States	Mean
Education	High	5 (4.50) 4	7 (6.00) 5	5.25
	Low	4 4 (3.67) 3	6 6 (6.00) 6	4.83
	Mean	4.00	6.00	5.00

In symbols:

		Germany	United States	Mean
	High	y_{111} y_{112} $(\bar{y}_{11})$	y_{121} y_{122} $(\bar{y}_{12})$	$\bar{y}_{1.}$
	Low	y_{211} y_{212} $(\bar{y}_{21})$ y_{213}	y_{221} y_{222} $(\bar{y}_{22})$ y_{223}	$\bar{y}_{2.}$
	Mean	$\bar{y}_{.1}$	$\bar{y}_{.2}$	$\bar{y}$

produces a higher mean subjective competence than the one produced by a low education, but in the United States the two corresponding means do not show that same difference. What we have here is an example of what is known as an interaction effect. There is something special about a high education in Germany that cannot simply be explained by the effect of a high education plus the effect of living in Germany; and similarly for the other three cells in the table. We return to further discussions of the interaction effect below.

Relationship among the explanatory variables. Before we can formally define and examine the interaction effect we need to turn to another aspect of two-day analysis of variance. This has to do with the relationship between the two explanatory variables themselves, here the relationship between country and education. This relationship is examined by considering the contingency table that is formed by counting the number of observations in each cell. This contingency table is seen in table 9.

The chi-square value for the frequencies in table 9 equals zero, as does any of the correlation coefficients we might use to measure the strength of a relationship between two nominal level variables. For the data in

TABLE 9
Contingency Table Showing the Frequences for
the Data in Table 8

		Country		
		Germany	United States	Total
Educa-tion	High	2	2	4
	Low	3	3	6
	Total	5	5	10

In symbols:

	Germany	United States	Total
High	n_{11}	n_{12}	$n_{1.}$
Low	n_{21}	n_{22}	$n_{2.}$
Total	$n_{.1}$	$n_{.2}$	n

table 9 there is therefore no relationship between the two variables. It should be emphasized that the chi-square value computed here has nothing to do with the possible effects of the explanatory variables on the dependent variable. The chi-square analysis concerns only the possible relationship between the explanatory variables themselves.

Establishing whether the two explanatory variables in an analysis of variance are themselves correlated is important. It is important because when the two explanatory variables are unrelated, as they are in this example, then it is possible to assess the unique effect of each of them on the dependent variable. But when the two explanatory variables are correlated with each other, then it is not possible to disentangle their separate effects on the dependent variable. In this example the explanatory variables are uncorrelated, and in the example below with the Almond and Verba data the explanatory variables are correlated.

The same problem of correlated explanatory variables occurs in multiple regression analysis, where it is labeled the problem of multicollinearity.[4] That name is not commonly used in connection with analysis of variance. In analysis of variance there are usually separate presentations for the case where the cell frequencies are equal or proportional, as here, and for the case when the cell frequencies are unequal and multicollinearity is present.

The existence of multicollinearity when doing analysis of variance is best established by the computation of chi-square for the frequency table formed by the two explanatory variables. The chi-square value equals zero when the cell frequencies are equal or when the frequencies in the rows (columns) are multiples of the frequencies in the other rows (columns). In our case the frequencies in the second row are 1.5 times the frequencies in the first row, and chi-square equals zero. In the second half of this chapter we deal with the case where the explanatory variables are not uncorrelated.

The problem of multicollinearity in analysis of variance is one reason why these methods have been used more often in experimental studies than in observational studies. When one is dealing with experimental research one can assign equal numbers of observations for each combination of categories for the explanatory variables. With observational studies one does not have the same freedom. Before collecting the data in large surveys there is no way of knowing what the cell frequencies will be; that is, how many Germans with a high education, how many Germans with a low education, and so on, one would get.

Effect of both variables. In order to understand better how a two-way analysis of variance is performed one can break the manipulations into several stages. As a first step the total combined effect of the two explanatory variables can be measured. This is done by stringing out the data according to a combined variable, in our case a combined country-education variable, and measuring the effect of this combined variable using the methods from the one-way analysis of variance. For our example the combined variable leads to four groups of observations. The data for this analysis and the results are shown in table 10.

The sum of squares for country and education together equals 10.83 and the residual sum of squares equals 3.17, resulting in an F-value of 6.83. This value of F is significant at the 0.023 level; that is, the probability of getting an F-value this large or larger when the two variables have no effect on the dependent variable equals 0.023.

In the general case with two explanatory variables A and B, where A has r categories and B has c categories, the number of categories for the combined A and B variable equals rc. Let n_{ij} denote the number of observations in the cell defined by the i-th category of A and the j-th category

TABLE 10
Subjective Competence Scores for Country and Education
Analyzed as a One-Way Analysis of Variance

	High educ. Germany	High educ. U.S.	Low educ. Germany	Low educ. U.S.
	5	7	4	6
	4	5	4	6
			3	6
Mean	4.50	6.00	3.67	6.00

Overall mean: 5.00

Source	Sum of squares	Degrees of freedom	Mean squares	F ratio	Significance
Country and Education	10.83	3	3.61	6.83	0.023
Residual	3.17	6	0.53		
Total	14.00	9			

$E^2 = 0.77$

for B. The sum of squares for the combined A and B variable is then found according to the expression

$$SS \text{ A and B} = \Sigma n_{ij}(\bar{y}_{ij} - \bar{y})^2$$

where the summation is across all the cells. The total sum of squares is found, as before, by subtracting the overall mean from each observation, squaring these differences, and adding them all up. Finally, the residual sum of squares is found as the difference between the total sum of squares and the sum of squares due to the combined A and B variable.

Three separate effects. The next step consists of decomposing the sum of squares for the combined variable into separate parts. In our example the combined country and education variable has three degrees of freedom for its sum of squares, and each component will therefore have one degree of freedom. The underlying mathematical theory states that each of these components of the sum of squares for the combined variable will result in an F-ratio when divided by the residual mean square, which here equals 0.53.

We already have one of the components we seek. This is the one from the one-way analysis of variance in chapter 2, where the effect of the country variable is investigated. It is the sum of squares for the country variable, which we found is equal to 10.00 on one degree of freedom for these data. Similarly, the second component is the sum of squares due to the education variable. This sum of squares can be found by performing a one-way analysis of variance with education as the explanatory variable.

However, it is not necessary to perform the complete analysis; all we need is the sum of squares due to education. From table 8 we have the mean subjective competence score for the four observations with a high education equal to 5.25, and the mean for the six observations with a low education equal to 4.83. The overall mean equals 5.00, and the sum of squares for education therefore becomes

$$4(5.25 - 5.00)^2 + 6(4.83 - 5.00)^2 = 0.42$$

on one degree of freedom.

By now we have found two of the three components. Of an overall sum of squares of 10.83 on three degrees of freedom due to the combined effect of country and education, 10.00 on one degree of freedom is due to the country variable and 0.42 on one degree is due to the education variable. That leaves a sum of squares of

$$10.83 - 10.00 - 0.42 = 0.41$$

on $3 - 1 - 1 = 1$ degrees of freedom. This sum of squares measures the effect of country and education on subjective competence, over and beyond the separate effects of country and education.

In the general case with r rows and c columns the sum of squares for the combined A and B variable is also decomposed into three components. The sum of squares for the A variable is found from the expression

$$SSA = \Sigma n_{i\cdot} (\bar{y}_{i\cdot} - \bar{y})^2 \qquad r - 1 \text{ d.f.}$$

where the summation is across all r categories of the A variable. Similarly, the sum of squares for the B variable is found from the expression

$$SSB = \Sigma n_{\cdot j} (\bar{y}_{\cdot j} - \bar{y})^2 \qquad c - 1 \text{ d.f.}$$

where the summation is across all c categories of the B variable. The sum of squares for the AB interaction variable is found by subtracting the sums of squares for A and B from the sum of squares for the combined A and B variable. That is,

$$SSInteraction = SS \text{ A and B} - SSA - SSB$$

on $(rc - 1) - (r - 1) - (c - 1) = (r - 1)(c - 1)$ degrees of freedom. In our example $r = c = 2$ and the degrees of freedom for the sum of squares for interaction equals one, as it should.

Analysis of variance table. The various results for this example are summarized in table 11 which lists the various sums of squares, degrees of freedom, mean squares, and F-ratios. The education and interaction F-ratios are both less than 1.00 and are therefore not very significant. But the country effect is significant, and because of the large value of the correlation ratio (0.71) the effect is also strong.

The country effect has been established again, even though not as convincingly here as in chapter 2. There the residual sum of squares was equal to 4.00 on 8 degrees of freedom, resulting in a residual mean square of 0.50. Here the sums of squares due to education and the interaction have been taken out of the old residual sum of squares, decreasing the residual sum of squares from 4.00 to 3.17. But at the same time the residual degrees of freedom decreased from 8 to 6, and the residual mean square becomes $3.17/6 = 0.53$. Thus, the new residual mean square is actually

TABLE 11
Analysis of Variance Table for the Data in Table 8

Source	Sum of squares	Degrees of freedom	Mean squares	F-ratio	Signi-ficance
Country	10.00	1	10.00	18.87	0.005
Education	0.42	1	0.42	0.79	0.40
Interaction	0.41	1	0.41	0.77	0.41
Residual	3.17	6	0.53		
Total	14.00	9			

E^2 (country) = 0.71 E^2 (educ) = 0.03 E^2 (int) = 0.03

larger for the two-way analysis than for the one-way analysis, resulting in a smaller F-value for the country variable. The F-value decreased from 20.00 to 18.87.

The main reason for the decrease in the F-values is that the example is based on a very small number of observations. With 10 times as many observations, for example, the F-value would have increased instead of decreased. With 100 observations the residual sum of squares of 4.00 would have been based on 98 degrees of freedom for the one-way analysis of variance. Then the residual mean square would have been 4.00/98 = 0.041 and F = 10.00/0.041 = 245.00. For the two-way analysis the residual sum of squares of 3.17 would have been based on 96 degrees of freedom, and the residual mean square would therefore have been 3.17/96 = 0.033 and F = 10.00/0.033 = 302.84.

With that many observations and all the other numbers staying the same, the reduction in residual degrees of freedom from 98 to 96 is more than offset by the reduction in residual sum of squares from 4.00 to 3.17. These changes result in an increase in the F-value from 245.00 to 302.84. This type of increase is much more typical of what happens when one goes from a one-way analysis of variance to a two-way analysis.

Effects of being in a particular row, column, and cell. The overall effects of the variables are shown in the sum of squares in table 11. The country variable seems to be most important with a contribution of 10.00 to the total sum of squares, the education variable contributes 0.42, the interaction of country and education contributes 0.41, and all other variables contribute 3.17. But these numbers are overall measures, and they do not explain why it is that a particular observation has the value that it does. Instead, we want numbers that measure the effect of each category of the variables.

The first value in the upper left cell of table 8 equals 5.00. That value equals 5.00 because that person has been affected by four variables:

(1) the high category of the education variable,
(2) the Germany category of the country variable,
(3) the high-Germany category of the education-country variable, and
(4) the residual variable, that is, the joint effect of all other variables.

Each of the 10 people in this example has been affected by these four variables, and we see the presence of the effects from the fact that the values of the observations differ from each other. Had the observations all been the same we would not have been able to conclude that any effects are present. Thus, we are more interested in the variation in the scores than their absolute level. The scores vary because it seems to make a difference for subjective competence whether one lives in one country versus another, whether one has a high versus a low education, and so forth.

We would like to measure the effects of these variables for each of the categories for each variable. We postulate that for a particular observation the four effects are additive, which means that we get each of the 10 scores in table 8 by adding up four effects. One way to get an understanding of how the effects operate is to subtract from the observed values successive numbers that can be interpreted as the various effects. But all effects are measured as deviations from the overall mean $\bar{y}$ (=5.00), and we therefore subtract 5.00 from each of the observations. The resulting scores are shown in table 12.

The four types of effects are still present. This can be seen from the fact that most of the observations and the means are still different. First we consider the effect of the education variable. The mean of the high education scores equals 0.250, and we interpret that to say that a value of 0.250 has been added to the subjective competence score for each person who receives a high education. Thus, the effect of a high education can be measured to equal 0.250. Similarly, a low education amounts to a subtraction of 0.167, and the effect of a low education can be measured to equal −0.167.

In symbols, let $\hat{\alpha}_1$ denote the estimated effect of being in the first row (high education) and $\hat{\alpha}_2$ the effect of being in the second row (low education). We then have

TABLE 12

Subjective Competence Scores After the Mean $\bar{y} = 5.000$ Has Been Subtracted from Each Observation

	Germany	U.S.	Mean
High	0.000 −1.000 (−0.500)	2.000 0.000 (1.000)	0.250
Low	−1.000 −1.000 −2.000 (−1.333)	1.000 1.000 1.000 (1.000)	−0.167
Mean	−1.000	1.000	0.000

$$\hat{\alpha}_1 = \bar{y}_1. - \bar{y} = 5.250 - 5.000 = 0.250$$

$$\hat{\alpha}_2 = \bar{y}_2. - \bar{y} = 4.833 - 5.000 = -0.167$$

where $\bar{y}_1.$ and $\bar{y}_2.$ are the original means of rows 1 and 2, respectively. We note that the two effects of education have different sizes and that $4(0.250) + 6(-0.167) = 0$, where 4 and 6 are the number of observations in the two corresponding rows. In general, the effect of the i-th category of the row variable is estimated as

$$\hat{\alpha}_i = \bar{y}_i. - \bar{y}$$

When each estimated effect is multiplied by the number of observations in the same row, then the sum of these products is always equal to zero.

Now that the effects of high and low educations have been identified, we can subtract out these effects and see if any variations in the resulting scores remain to be explained. Subtracting 0.250 from all the observations in the first row and −0.167 from all the observations in the second row of table 3.5 we get the results in table 13.

TABLE 13
Subjective Competence Scores After the Overall Mean and the
Row Effects Have Been Subtracted from Each Observation

	Germany	U.S.	Mean
High	-0.250 -1.250 (-0.750)	1.750 -0.250 (0.750)	0.000
Low	-0.833 -0.833 -1.833 (-1.167)	1.167 1.167 1.167 (1.167)	0.000
Mean	-1.000	1.000	0.000

The two row means are now the same and equal to zero because the effects of education have been removed from the observations. But the four cell means are still different, as are the two column means. Since the column mean for Germany equals -1.000, the effect of being a German amounts to having a value of 1.000 subtracted from one's score. Similarly, the effect of living in the United States is to have a value of 1.000 added to one's score. Thus, the effects of being in columns 1 and 2 can be measured by

$$\hat{\beta}_1 = \bar{y}_{.1} - \bar{y} = 4.000 - 5.000 = -1.000$$

$$\hat{\beta}_2 = \bar{y}_{.2} - \bar{y} = 6.000 - 5.000 = 1.000$$

where $\bar{y}_{.1}$ and $\bar{y}_{.2}$ are the original means in columns 1 and 2, respectively. In general, the effect of being in the j-th category of the column variable is estimated to equal

$$\hat{\beta}_j = \bar{y}_{.j} - \bar{y}$$

Next we subtract the effect of living in Germany from all the observations in the first column and the effect of living in the United States from all the observations in the second column to see if there are any further effects present in these data. The results are shown in table 14.

Now we have subtracted out from each observation both the effect of education and the effect of country. What is left of the observations must therefore be due to the two remaining variables, the interaction variable and the residual variable. The four cell means give us clues to the magnitude of the effects of the interaction variable. The mean of a high education in Germany still equals 0.250 after we have subtracted out both the effect of a high education and the effect of living in Germany. The effect of the country-education interaction variable for a high education in Germany is therefore estimated to equal 0.250. Similarly, the effect of the interaction variable for a high education in the United States equals −0.250, the interaction effect for a low education in Germany equals −0.167, and the interaction effect for a low education in the United States equals 0.167.

In symbols the four interaction effects can be expressed as follows:

High education in Germany:

$$\hat{\gamma}_{11} = 4.500 - 5.000 - 0.250 - (-1.000)$$

$$= \bar{y}_{11} - \hat{\mu} - \hat{\alpha}_1 - \hat{\beta}_1$$

$$= \bar{y}_{11} - \bar{y} - (\bar{y}_{1.} - \bar{y}) - (\bar{y}_{.1} - \bar{y})$$

$$= \bar{y}_{11} - \bar{y}_{1.} - \bar{y}_{.1} + \bar{y}$$

$$= 4.500 - 5.250 - 4.000 + 5.000$$

$$= 0.250$$

High education in the United States:

$$\hat{\gamma}_{12} = y_{12} - \hat{\mu} - \hat{\alpha}_1 - \hat{\beta}_2$$

$$= \bar{y}_{12} - \bar{y}_{1.} - \bar{y}_{.2} + \bar{y}$$

$$= -0.250$$

Low education in Germany:

$$\hat{\gamma}_{21} = \bar{y}_{21} - \hat{\mu} - \hat{\alpha}_2 - \hat{\beta}_1$$

$$= \bar{y}_{21} - \bar{y}_{2.} - \bar{y}_{.1} + \bar{y}$$

$$= -0.167$$

Low education in the United States:

$$\hat{\gamma}_{22} = \bar{y}_{22} - \hat{\mu} - \hat{\alpha}_2 - \hat{\beta}_2$$

$$= \bar{y}_{22} - \bar{y}_{2.} - \bar{y}_{.2} + \bar{y}$$

$$= 0.167$$

TABLE 14
Subjective Competence Scores After the Overall Mean, Row Effects
and Column Effects Have Been Subtracted from Each Observation

	Germany	U.S.	Mean
High	0.750	0.750	0.000
	-0.250 (0.250)	-1.250 (-0.250)	
Low	0.167	0.167	
	0.167	0.167	0.000
	-0.833 (-0.167)	0.167 (0.167)	
Mean	0.000	0.000	0.000

In general, the estimate of the interaction effect for the cell located in the i-th row and j-th column of the table is found by taking the cell mean, subtracting the mean of the row in which the cell is located, subtracting the mean of the column in which the cell is located and, finally, adding the overall mean. That gives us the formula

$$\hat{\gamma}_{ij} = \overline{y}_{ij} - \overline{y}_{i \cdot} - \overline{y}_{\cdot j} + \overline{y}$$

The residual variable. Finally we can subtract the interaction effects from the remaining parts of the observations. The results are given in table 15. The only thing left of the original observed values by now are the parts due to the residual variable. The effect of the residual variable on the first person in the cell for a high education in Germany equals 0.500, and similarly for the nine other people.

From the way the various effects have been subtracted out we can see from table 15 that the residuals are obtained as deviations of the original observations in table 8 from the mean of the observations in the corresponding cell. This is not surprising, because if the only variables operating here were the country, the education, and the country-education interaction variables, then all the observations in a particular cell would be equal. The extent to which the observations in a cell differ from each other must be due to the residual variable.

TABLE 15

Subjective Competence Scores After the Overall Mean, Row Effects, Column Effects and Interaction Effects Have Been Subtracted from the Observations, Giving the Values of the Residual Variable

	Germany	U.S.	Mean
High	0.500 -0.500	1.000 -1.000	0.000
Low	0.333 0.333 -0.666	0.000 0.000 0.000	0.000
Mean	0.000	0.000	0.000

One of the formal assumptions of analysis of variance is again that the residuals form a normal distribution. The numbers we have here are only the estimated values of the residual variable, but their distribution should not deviate too much from the normal distribution. Figure 6 shows the frequency distribution of the residuals we have here. The distribution is somewhat skewed, but it does not offer any strong evidence against normality. The assumption about normality is needed for the F-tests. Experience has shown that the F-tests are not adversely affected by deviations from normality in the distribution of the residuals. But even though the F-tests have been found to be robust in this sense, one should still examine the distribution of the residuals for normality. One should watch for skewness and single observations that lie far away from the rest of the observations. For a further discussion of the analysis of residuals see Anscombe and Tukey (1963) and Box (1953).

Sum of squares from effects. One way to measure the overall effect of the residual variable is to square each residual and add up these squares. The mean of the residuals is always equal to zero and gives no indication of the magnitudes of the residuals. The sum of squared residuals becomes

$$(0.500)^2 + (-0.500)^2 + \ldots + (0.000)^2 = 3.17$$

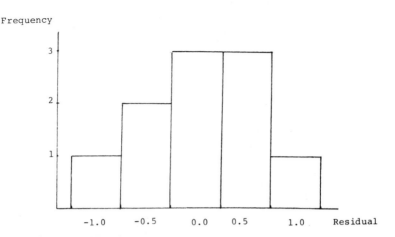

Figure 6: Frequency Distribution of the Residuals in Table 15

This is the very same sum of squares for the residual variable as found in table 11.

The variance of the residuals is found by dividing this sum by the appropriate number of degrees of freedom, here 6 and in general n-rc, where n is the total number of observations, r the number of rows and c the number of columns. In this case the variance becomes $s_e^2 = 3.17/6 = 0.53$ and the standard deviation $s_e = 0.73$. Thus, an average observation deviates 0.73 from the mean in the cell in which it is located.

The other sums of squares in table 11 can also be found from the effects that are identified above. For education we found that the effect on an observation of a high education is $\hat{\alpha}_1 = 0.250$ and the effect of a low education is $\hat{\alpha}_2 = -0.167$. Four people have a high education and six people have a low education. The total effect of education can then be measured by the sum of squares

$$4(0.250)^2 + 6(-0.167)^2 = 0.42$$

For the country variable there are five people from Germany with $\hat{\beta}_1 = -1.000$ and five people from the United States with $\hat{\beta}_2 = 1.000$. The overall effect of country can then be measured by the sum of squares

$$5(1.000)^2 + 5(-1.000)^2 = 10.00$$

Finally, the interaction sum of squares in table 11 equals 0.41. That sum of squares can also be found from the four interaction effects

$$\hat{\gamma}_{11} = 0.250 \quad \hat{\gamma}_{12} = -0.250 \quad \hat{\gamma}_{21} = -0.167 \quad \hat{\gamma}_{22} = 0.167$$

by squaring each of the terms, multiplying each square by the number of observations in the corresponding cell and adding. That is,

$$2(0.250)^2 + 2(-0.250)^2 + 3(0.167)^2 + 3(-0.167)^2 = 0.417$$

Formal model. This discussion started by focusing on the observation with value equal to 5.000 in the cell for high education in Germany. We are now able to decompose that value, using the effects that have been identified above. We have, in words:

> Observed
> value   = overall mean
>   + effect of high education
>   + effect of Germany
>   + effect of a high education in Germany
>   + effect of all other variables

In numbers:

$$5.000 = 5.000 + 0.250 - 1.000 + 0.250 + 0.500$$

and similarly for each of the other nine observations. The magnitudes of the effects help us analyze the relative importance of the explanatory variables on the dependent variable both directly for each of the categories of the explanatory variables and in an overall way for each variable by the corresponding sums of squares.

The formal model that underlies a two-way analysis of variance specifies that the k-th observation in the cell located in the i-th row and j-th column in a table like table 9, y_{ijk}, can be written as a sum

$$y_{ijk} = \mu + \alpha_i + \beta_j + \gamma_{ij} + \epsilon_{ijk}$$

where μ is a constant,

    α_i is the effect of the i-th category of the row variable,

    β_j is the effect of the j-th category of the column variable,

γ_{ij} is the effect of the (i,j)-th category of the interaction variable, and

ϵ_{ijk} is the effect of all other variables.

The expressions we have for these quantities above are estimators of the model parameters, and the numerical estimates are obtained from our particular sample. The estimates are all found using the criterion that the sum of squared residuals should be as small as possible.

Hypothesis testing. Because we have only estimated effects, we want to use the sample data as well as we can to conclude about the corresponding true effects in the larger population from which the sample came. This is done by means of testing a series of hypotheses about the population effects. In our example one hypothesis is that education has no effect, that is,

$$H_0: \alpha_1 = \alpha_2 (=0)$$

Because the effects are measured as deviations from the overall mean some will be negative and some will be positive. The weighted sum of the effects, where the weights are the number of observations in each category, is equal to zero. Thus, the only way the effects can be equal is for the effects all to equal zero. Another way of stating the same hypothesis is that the population means of the dependent variable are alike for all categories of the row variable.

This hypothesis is tested by the F-value for education in the analysis of variance table in table 11. There the F-value on the line for education is an estimate of the quantity

$$1.00 + (4\alpha_1^2 + 6\alpha_2^2)/\sigma_\epsilon^2$$

where σ_ϵ^2 is the variance of the residuals. If the hypothesis is true, that is, the α's really are equal to zero, then the value of this quantity equals 1.00. The observed sample value of F would not be exactly equal to 1.00, but it will be fairly close.

On the other hand, when the hypothesis is not true, then the value of the quantity above will be larger than 1.00, and the sample value of F will normally be a good deal larger than 1.00. Thus, a small value of F will lead us to think that the hypothesis of no effects is true and a large value of F will lead us to think that the hypothesis is false. The specific cut-off point for F for which we reject the hypothesis is found in tables of the F-distribution. In our example, where there is one degree of freedom for education and six degrees of freedom for the residuals, we reject the hypothesis if the observed value of F is larger than 5.99, with a 5% significance level.

In the general case, with r rows in the table, we can test the hypothesis that there are no differences between the rows in the same way. The F-value is computed as in our example, and it is an estimate of the quantity

$$1.00 + (n_1\alpha_1^2 + n_2\alpha_2^2 + \ldots + n_r\alpha_r^2)/\sigma_\epsilon^2(r-1)$$

The hypothesis of no differences in the row means is rejected when the observed F-value is larger than the value obtained from the F-table, with $r-1$ and $n-rc$ degrees of freedom. The n's above are the number of observations in the various rows.

The hypothesis that there is no difference between the categories that form the columns, here that there is no difference between the two countries, is tested the same way. With c columns the hypothesis states

$$H_0: \beta_1 = \beta_2 = \ldots = \beta_c (=0)$$

It is tested by the corresponding F-value, which is an estimate of the quantity

$$1.00 + (n_1\beta_1^2 + n_2\beta_2^2 + \ldots + n_c\beta_c^2)/\sigma_\epsilon^2(c-1)$$

where the n's here are the number of observations in the various columns. If the observed value of F is a good deal larger than 1.00, then the hypothesis is rejected. The critical value is found from tables of the F-distribution, using $c-1$ and $n-rc$ degrees of freedom. Rejecting the hypothesis means that we conclude that at least some of the β's are different from zero. In our example F equals 18.87, and the hypothesis is rejected.

Finally, it is possible to test whether the interaction effects all equal zero. The hypothesis states that they are all equal, meaning that they all equal zero. In our example,

$$H_0: \gamma_{11} = \gamma_{12} = \gamma_{21} = \gamma_{22} (=0)$$

The hypothesis is tested by the F-ratio for interaction, because that F is an estimate of the quantity

$$1.00 + (n_{11}\gamma_{11}^2 + n_{12}\gamma_{12}^2 + n_{21}\gamma_{21}^2 + n_{22}\gamma_{22}^2)/\sigma_\epsilon^2(2-1)(2-1)$$

With F = 0.41 there is no reason to reject the hypothesis of zero interaction effects. To find the value of F necessary to reject the hypothesis we consult a table for the F-distribution, using $(r-1)(c-1)$ and $n-rc$ degrees of freedom.

If the test for interaction reveals no significance, the formal model can take that into account and be rewritten as the equation

$$y_{ijk} = \mu + \alpha_i + \beta_j + \epsilon_{ijk}$$

Here there are no interaction terms, and this is a simpler, additive model. Any observation is determined only by the effect of being in a particular row plus the effect of being in a particular column, plus the effect of the residual variable. In such a case the residual sum of squares is recomputed, as discussed in the section called pooling nonsignificant sums of squares.

The interpretation of the data is more troublesome if one or perhaps both main effects, row and column effects, are not significant while the interaction effect is significant. Suppose the interaction effect had been significant here. The effect of education is not significant, and there would therefore be good reasons to drop this variable from the analysis and further considerations. But because of the significant interaction effect we cannot remove the education variable. If we took away the education variable we would be taking away the interaction variable at the same time, and it does not make sense to eliminate a variable we had found to have a significant effect.

One also has to keep in mind that even a very large and significant F-value only tells us that the corresponding effects differ from zero. It does not tell us anything about how much the effects differ from zero, and that is usually what is of interest. The magnitude of the differences can only be established by computing the effects for each category and interpreting them relative to the substantive topic under study.

Related Explanatory Variables

Related variables. With observational studies, as opposed to experimental studies, one has little or no control over how many observations one will have in the various categories of the variables. In particular, with two explanatory variables and a dependent variable, it may well be that the two explanatory variables themselves are related because of unequal frequencies of observations in the various cells. When this is the case, it becomes more difficult to study the effects of the variables involved. This is because the influence of a particular explanatory variable now can take two paths. It influences the dependent variable directly, but, in addition, it also influences the dependent variable indirectly through the other explanatory variable.

The number of observations in the various cells differ greatly in the Almond and Verba study, as shown in table 16. The frequencies in table 16 are less than those used in the one-way analysis because of the problem of missing values. Neither the cell entries nor the percentages for the three educational levels are equal in the five countries. The two explanatory variables—country and education—are therefore correlated, and the data in table 16 give a chi-square value of 884.36, which is significant far beyond the 0.001 level on 6 degrees of freedom for the relationship between the two explanatory variables themselves. For the purposes of analysis of variance, however, it matters not so much that the value is significant, but that it is non-zero.

The way in which such data can be analyzed using a two-way analysis of variance model is demonstrated by a small illustrative example involving two countries and two levels of education. The data for this example are shown in table 17. We have taken the data from the previous section and changed the education for one respondent in the United States from low to high. With this change the analysis in the previous section has no rele-

TABLE 16

Number of Observations in Each Cell and Row Percentages, Showing the Relationship Between Country and Education for Data from the Five Nation Study by Almond and Verba

| | Education | | | |
Country	Grade school	High school	College	Total
Great Britain	559 (62.5%)	312 (34.9)	24 (2.7)	895 (100.1%)
Italy	619 (81.7%)	89 (11.7)	50 (6.6)	758 (100.0%)
Mexico	703 (85.3%)	100 (12.1)	31 (2.5)	824 (99.9%)
United States	321 (34.1%)	433 (46.1)	186 (19.8)	940 (100.0%)
West Germany	713 (83.1%)	120 (14.0)	25 (2.9)	895 (100.0%)
Total	2915 (68.2%)	1054 (24.6)	306 (7.2)	4275 (100.0%)

TABLE 17
Subjective Competence Scores Classified by Country and Education,
with the Country and Education Variables Related

	Germany	U.S.	Mean
High educ.	5 4 (4.50)	5 6 7 (6.00)	5.40
Low educ.	4 4 (3.67) 3	6 6 (6.00)	4.60
Mean	4.00	6.00	5.00

Relationship between education and country

	Germany	U.S.	Total
High educ	2	3	5
Low educ	3	2	5
Total	5	5	10

$\phi = 0.2$ $X^2 = 0.4$

vance any more, and the data have to be reanalyzed. The relationship between education and country can be studied by examining how many observations there are in the four cells that make up the observations for the
dependent variable in table 17. For this example the distribution of the
frequencies result in a correlation coefficient $\phi = 0.2$ and a chi-square value

of 0.40. The small value of chi-square is mainly due to the few observations, and the main point is that with a correlation coefficient as large as 0.2 we clearly have a relationship between the two variables.

In addition to the country and education variables, we still have the country-education interaction variable as the joint effect of country and education on subjective competence, over and beyond their separate effects. The two main variables now are correlated, and the interaction variable is usually correlated with both of the two main variables. The various sums of squares are computed by letting the two main variables explain as much of the variation in the dependent variable as they can and allocating the remaining part of the combined sum of squares for the main variables and interaction variable to the interaction variable.

Formal model and plan for analysis. The formal model for two-way analysis of variance, when the two nominal variables are correlated, is the same as when they are not correlated. A particular observation y_{ijk}, the k-th observation in the cell defined by the i-th row and the j-th column, is seen as made up of the same type of components as before. The model can therefore be written as the equation

$$y_{ijk} = \mu + \alpha_i + \beta_j + \gamma_{ij} + \epsilon_{ijk}$$

where the components on the right hand side of the equation are defined in the previous section.

The effects of the three variables and the residual variable are measured by sums of squares, as before, and the significance of the effects of the variables can be decided by various F-ratios. With the aid of a computer one can find the appropriate sums of squares directly, but in order to gain an understanding of how the model applies to data of the kind we have here, we give a fairly detailed account of how the computations are made. The first step again consists of getting the residual sum of squares and a sum of squares that represents the combined effect of the two main variables and the interaction variable. This is done by stringing out the cells in the table with r rows and c columns into a one-way analysis of variance with rc groups. This analysis gives the total sum of squares TSS, the residual sum of squares RSS and a sum of squares TSS-RSS representing the combined effect of the two main variables and the interaction variable.

One can proceed from this point in a couple of directions. One direction consists of setting up appropriate dummy variables and doing what amounts to a regression analysis which gives a sum of squares that is due to the combined effect of the two main variables. By subtracting that sum from the sum TSS-RSS obtained above, one gets the sum of squares due to the interaction variable. Next, one can do a one-way analysis for one of the two main variables and get the sum of squares measuring the effect of that variable. Finally, the sum of squares for the other main variable is obtained by subtraction from the sum obtained by the regression analysis.

The other way of doing the analysis is the way it is done below, using the data introduced in table 17. First, the sum of squares for education is found directly from a one-way analysis of variance. Next, the sum of squares for country, after education has been allowed to explain as much of the total sum of squares as it can, is found from a formula we give without any proof. From the various sums of squares computed by now it is possible to find the interaction sum of squares by subtraction. We also turn the analysis around and first find the country sum of squares, and then let the education variable explain as much as it can of the remaining variation. Due to the correlation between the country and education variables the results are different in the two cases.

Sums of squares. Even though we have changed the data in this section by moving an observation from one cell to another, it so happens that a one-way analysis of variance of the resulting four groups of observations gives the same results as we had when we did not have correlated explanatory variables for our particular example. The results are the same as those displayed in table 10, where the total sum of squares equals 14.00, and the corresponding number of degrees of freedom is 9. The residual sum of squares equals 3.17, with 6 degrees of freedom. The sum of squares due to the combined effect of the country, education, and interaction variables therefore equals $14.00 - 3.17 = 10.83$, with 3 degrees of freedom.

The separate effect of education, as measured by the education sum of squares, can be found from the means in table 17. We have

$$
\begin{aligned}
\text{EducSS} &= n_1.(\bar{y}_1. - \bar{y})^2 + n_2.(\bar{y}_2. - \bar{y})^2 \\
&= 5(5.40 - 5.00)^2 + 5(4.60 - 5.00)^2 \\
&= 1.60
\end{aligned}
$$

The effect of country, after education has been allowed to explain as much as it can, is found from the expression

$$
\frac{(n._1\bar{y}._1 - n_{11}\bar{y}_1. - n_{21}\bar{y}_2.)^2}{n._1 - n_{11}^2/n_1. - n_{21}^2/n_2.}
$$

$$
= \frac{(5(4.00) - 2(5.40) - 3(4.60))^2}{5 - 4/5 - 9/5}
$$

$$
= 8.82 = \text{CounSS}
$$

The sum of squares for interaction can finally be found by subtraction. Country, education, and interaction together account for a sum of squares equal to 10.83. From the two sums above we get

$$\text{IntSS} = 10.83 - \text{EducSS} - \text{CounSS}$$
$$= 10.83 - 1.60 - 8.82 = 0.41$$

The results are summarized in table 18. That table also shows the results when the country sum of squares is computed directly from the country means through a one-way analysis of variance. The education sum of squares then shows the amount of variation that is left for the education variable to explain, after the country variable first has been allowed to explain all the variation it can.

TABLE 18

Analysis of Variance Tables for the Example with Correlated Explanatory Variables

Source	Sum of squares	Degrees of freedom	Mean squares	F ratio	Significance
Education	1.60	1	1.60	3.03	0.13
Country, after education	8.82	1	8.82	16.69	0.006
Interaction	0.41	1	0.41	0.78	0.41
Residual	3.17	6	0.53		
Total	14.00	9			

Source	Sum of squares	Degrees of freedom	Mean squares	F ratio	Significance
Country	10.00	1	10.00	18.93	0.005
Education, after country	0.42	1	0.42	0.79	0.41
Interaction	0.41	1	0.41	0.78	0.41
Residual	3.17	6	0.53		
Total	14.00	9			

The two results differ. The interaction sum of squares is always the same in the two cases, but the sums of squares for the two main variables differ. Country, for example, has a sum of squares of 10.00 when country is entered first. But when education first explains all it can explain, the remaining sum of squares for the country variable reduces to 8.82; and similarly for education. What matters, then, is the order in which the variables are introduced into the analysis.

Importance of order. This example shows the importance of the order in which the variables are introduced. The existing output for computer programs is not always as specific as it should be on this point, and it is not always clear from the output in what order the variables have been treated. For a further discussion of this point see Francis (1973).

The impact on the education variable is particularly severe in our example. When education is introduced first, the F-ratio is as large as 3.03, which with one and six degrees of freedom has a p-value (significance) of 0.13. But when the education variable is left to pick up after the country variable, the sum of squares drops from 1.60 to 0.42, and the corresponding F-value drops to 0.79, with a p-value as large as 0.41.

Results based on the full data set from the five nations also reveal sharp differences depending on which of the two explanatory variables is entered first. When country is entered first, its sum of squares amounts to 259.65, as shown in table 19. The table also shows that when country is entered after education, that figure is cut in half to 125.37. Similarly, the sum of squares for education is reduced from 332.43 to 198.15 when that variable is entered after the country variable. In both cases the reduction in the sum of squares from when the variable is entered first to when it is entered second equals 134.28.

One way to interpret that difference is to say that the number represents the part of the sum of squares that is shared by the two variables. Figure 7 shows these various numbers, using areas to represent sums of squares. The area in the middle represents the portion of the variation in the subjective competence scores that is shared by both variables. The two outside areas point to that part of the variation which each of them explains alone, after the other variable has been entered into the analysis first.

The reason why each of the two explanatory variables loses such a sizeable share of its sum of squares when it is entered after the other variable lies in the nature of the relationship between the two variables and subjective competence. This relationship is illustrated in table 20 showing the percentage with at least a high school degree and mean subjective competence score for each country. In that table we find that a country with a

TABLE 19
Analysis of Variance Tables for the Almond and Verba Data

Source	Sum of squares	Degrees of freedom	Mean squares	F ratio	Signi-ficance
Country	259.65	4	64.91	87.75	0.00
Education, after country	198.15	2	99.08	133.94	0.00
Interaction	20.73	8	2.59	3.50	0.0005
Residual	3150.98	4260	0.74		
Total	3629.51	4274			
Education	332.43	2	166.22	224.72	0.00
Country, after education	125.37	4	31.34	42.37	0.00
Interaction	20.73	8	2.59	3.50	0.0005
Residual	3150.98	4260	0.74		
Total	3269.51	4274			

high level of education also shows a high level of subjective competence. The higher a country's level of education the higher, on the average, is its subjective competence mean. Such a positive, albeit imperfect correlation, implies that the overall level of subjective competence in a country may be attributable in large part to its level of economic and social development as indexed by education. Our findings indicate that slightly over half of the variation in subjective competence which initially was credited to national differences must also be credited to level of development (education). By entering the country variable after the education variable we get an estimate of the effect of national differences on subjective competence which would be observable if the five countries were equal in education. With levels of education held constant the direct effect of national differences on subjective competence is greatly reduced, but it is not eliminated.

Computing. Our analysis was done on the program called MANOVA in the OSIRIS program package. It performs analysis of variance with one

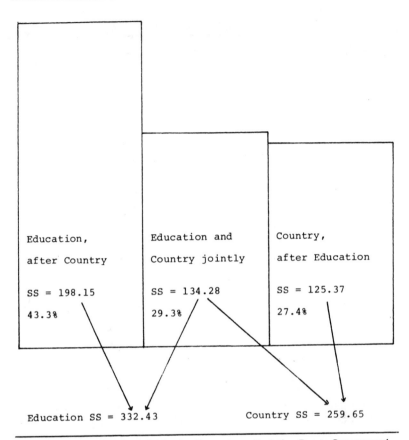

Figure 7: Partitioning of the Subjective Competence Variation Due to Country and Education (Exclusive of Interaction)

or more factor variables (predictor, control variable). It also performs multivariate analysis of variance, by which is meant the analysis of variance with more than one dependent variable. The name of the program, in fact, refers to this feature (Multiple Analysis of Variance). Since our discussion is confined to instances involving one and only one dependent variable, we ignore this feature of the program.

The output of this program presents to the novice a bewildering sequence of tables and matrices, but only two of them really matter as long as one is interested in questions involving just a single dependent variable. One of the tables is the output page headed *The Cell Means for Each Cell*. It provides the mean of the dependent variable as well as the number of cases for each cell of the analysis design. These are the data we show in

[68]

TABLE 20
Mean Subjective Competence Level and Level of Development (Education)

Country	Percent with at least a high school degree	Mean subjective competence score
United States	65.9	0.36
Great Britain	37.6	0.17
Italy	18.3	-0.33
West Germany	16.9	-0.21
Mexico	14.6	0.05

tables 7 and 16. The other table of interest is listed at the end of the MANOVA output. This table provides a standard analysis of variance table, with two exceptions. One is that the total sum of squares is not given. The other is an item called *Grand Mean* which appears in the table with the other items. Its sum of squares is calculated as $n\bar{y}^2$, where n represents the total number of cases and $\bar{y}$ the grand mean of the dependent variable. Note that in calculating the total sum of squares from the other sums of squares that are given, one should not include the sum of squares for the grand mean.

It is one of the important features of the MANOVA program that it allows for analysis of correlated explanatory variables, that is, designs with unequal as well as unproportional cell entries. In the event of correlated explanatory variables the program considers one of them first and allows it to pick up as much variation in the dependent variable as possible. The next explanatory variable is then allowed to pick up only variation which has not already been tapped by the first variable.

It is up to the investigator to decide which explanatory variable to introduce first, with its effect being unadjusted for the effects of the other explanatory variable, or variables in the event of more than two. The variable which is entered second will emerge with an effect which is adjusted for the effect of the first variable. In general, any variable is adjusted for the effect of preceding variables and unadjusted for the effects of variables which follow it.

It is the "variable list" card of the MANOVA setup where the investigator specifies the order of introduction. Paradoxically, one has to do it in a reverse order; that is, the variable placed last on the card is entered first in the analysis. Thus, the variable to be entered last should lead the list. One should note that this variable list card also includes the dependent variable. It always comes first, and then comes the string of explanatory variables. For further details one should consult the OSIRIS manual (1973).

Special Topics

One observation per cell. There are times when each combination of the explanatory variables has only one observation. That is, there is only one observation in each cell of the data table. This situation cannot be handled by the methods discussed so far. The two nominal level variables would be uncorrelated in this case, but another problem arises instead. It would not be possible to compute any F-ratios, and it would therefore be impossible to test the standard hypotheses about the presence of various effects.

It would not be possible to do any testing because we do not have any way of getting the denominator for the F-values. The denominator for the F's, as we have seen, is the mean square for the residuals, and it is obtained from the residual sum of squares. But each residual is obtained as the difference between the observed score and the mean in the corresponding cell of the table. With only one observation in the cell the mean is equal to that observation, and the residual would equal zero. With residuals equal to zero we would have the residual sum of squares equal to zero, and one cannot divide by zero.

The case with only one observation in each cell does not occur all that often, but this section is useful for another reason. If one wants to get some preliminary ideas of whether main effects are present or not without performing all the computations necessary for a complete analysis, one can first compute the mean of the dependent variable for the observations in each cell. By replacing the original observations with the cell means one then has the case where there is only one observation in each cell. These means can then be analyzed for the presence of main effects as outlined below.

The one way to do an analysis of variance with one observation in each cell is to make the assumption that there is no interaction effect present in the data. With such an assumption what ordinarily would have been the interaction sum of squares is now called the residual sum of squares and used in the denominator for the F-values. But we no longer have any way of testing for the presence of significant interaction effect, so we can only test for the presence of the two sets of main effects.

This also means that with unequal cell frequencies and correlated explanatory variables, it is possible to reduce the data to the case with one observation in each cell and proceed without having to worry about the order in which the variables are brought into the analysis. The one value we would use in each cell would be a central value like the mean.

In the small, numerical example with related explanatory variables we know the interaction effect is not significant. If we do the analysis over again, using the means in table 8 as single observations, we get the results shown in table 21. The means for the two educational categories, the two countries, and the overall mean are all changed because we now have the same number of observations in each of the four cells. There is also a large change in the analysis of variance table itself. There we see that the anal-

TABLE 21
Two-Way Analysis of Variance with One Observation per Cell

	Germany	US	Mean
High Educ	4.50	6.00	5.25
Low educ	3.67	6.00	4.83
Mean	4.08	6.00	5.04

$$\hat{\alpha}_1 = 5.25 - 5.04 = 0.21$$
$$\hat{\alpha}_2 = 4.83 - 5.04 = -0.21$$
$$\hat{\beta}_1 = 4.08 - 5.04 = -0.96$$
$$\hat{\beta}_2 = 6.00 - 5.04 = 0.96$$

Source	Sum of squares	Decrees of freedom	Mean squares	F-ratio	Significance
Country	3.67	1	3.67	21.17	0.14
Education	0.17	1	0.17	1.00	0.50
Residual	0.17	1	0.17		
Total	4.01	3			

ysis is now based on only three degrees of freedom. This means that we have lost a good deal of information relative to the analysis based on the original nine degrees of freedom. But in spite of this loss we find that the F-ratios for country and education are not all that different from the F-ratios we found earlier.

The analysis above is based on a nonsignificant interaction effect. Had we started with only one observation in each cell, it would not have been possible to determine whether or not the interaction effect is significant. The worst that could have happened would have been that there was a significant interaction effect present, unknown to us. What we here have taken as the residual sum of squares would then have been the effect of the residual variable plus the effect of the interaction variable. From statistical theory we must use only the residual variable in the denominator for the F-ratios. If there had been an interaction effect present as well, then the denominator for the F-ratios would have been too large and the F-values themselves would therefore have been too small. In other words, we may not be able to detect significant row and column effects even though they really are present. When we find a significant F-value, we therefore know that the significance level is on the conservative side.

Pooling nonsignificant sums of squares. Going back to table 11 we are in the situation where, in particular, the interaction variable does not have a significant effect. If we conclude from the small value of F for interaction that the interaction variable is not present in these data, then we are faced with the question of what produced the sum of squares of 0.41. One answer is that this sum of squares was produced by the random effect of all other variables that together with education and country work to determine the subjective competence scores. But we have already called that the effect of the residual variable, and we have measured the effect of the residual variable to have a sum of squares of 3.17. Now we have two measures of the residual effect. It is best to combine the two into one, because we get a larger number of degrees of freedom for the new sum of squares.

By combining the interaction sum of squares and the residual sum of squares we get a new residual sum of squares of $3.17 + 0.41 = 3.58$, with a total of seven degrees of freedom. That way the new residual mean square becomes $3.58/7 = 0.51$, and this is smaller than the old residual mean square. With a smaller residual mean square the F-values become larger, and that way we are better able to determine whether the two main variables really have effects that are present in the underlying population from which the sample data came.

The increases in the F-values in table 10 are not large in this case, but at the same time as the F-values get larger, the degrees of freedom for the denominator also get larger. This again means that the critical value of F for the rejection of the null hypothesis gets smaller. In table 11 we have an F-value of 18.87 with one and six degrees of freedom for the country variable. In order to reject the hypothesis of no country effect we need to have F larger than 5.99, with a 5% significance level. Using the new residual mean square the observed value of F increases to 19.60, and with one and seven degrees of freedom, the critical value of F has decreased to 5.59 according to the F-table.

Interaction as nonparallel lines. The interaction variable has been introduced as the joint effect of the two explanatory variables over and beyond the separate effects of each of the two explanatory variables. One way to get a better understanding of the presence of interaction in a set of data is to draw a picture like figure 8 for the data in our small example. The picture is made in the following way. The categories for one of the explanatory variables, and it makes no difference which of the two is chosen, are marked off along the horizontal axis. The dependent variable is marked off along the vertical axis, and then each of the cell means is plotted in its appropriate place. The figure shows the four cell means from the data in table 8. The final step in creating the figure consists of connecting the means that have the same category of the second explanatory variable. Here the two high education means are 4.50 and 6.00, and they are connected by the line marked high education, and similarly for the two means for low education.

The two lines in figure 8 are not parallel, and because they are not there is some interaction present in these data. The fact that the lines are not parallel means that it makes a difference for the dependent variable which country one is in if there is a change in education. The figure shows that in Germany the mean subjective competence changes from 3.67 to 4.50 when education changes from low to high. But in the United States there is no such change in the means when the level of education changes. In other words, it makes a difference for subjective competence what country one is in when education changes from low to high. This is because of the interaction effect.

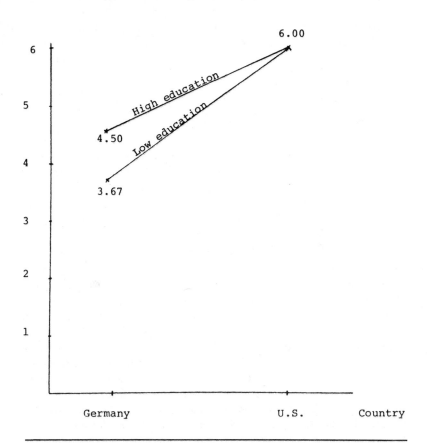

Figure 8: Means in Each of the Four Groups, Connected by Lines Showing the Presence of Interaction

4. ANALYSIS OF VARIANCE, SAMPLE OF CATEGORIES

One-Way Analysis

Sample of categories. The type of analysis of variance outlined in chapters 2 and 3 is called "model I" type of analysis. This model is also known as a "fixed effect" model. We now turn to "model II," also called the "random effect" model. This model, unlike the previous one, does not require observations on the dependent variable for each of the different categories of the explanatory variable or variables. Instead, model II handles situations where data are available for only a sample of categories.

The five nation example used in the previous two chapters can be stretched somewhat to illustrate the difference between the two types of models. Suppose that the investigators of that comparative study were interested specifically in those five countries and wanted their conclusions, drawn by methods of statistical inference, to apply to those five countries only; then the data should be analyzed according to what has been called model I type of analysis. On the other hand, one might want to reach conclusions, drawn by methods of statistical inference, which would apply to all 150 or so countries without requiring data on subjective competence from all the 150 countries. With that goal in mind, one would first draw a random sample of countries and then, within each of the chosen countries, draw a random sample of respondents. Data selected that way and for the purpose indicated should be analyzed according to the type II model. The conclusions we can draw from this analysis would apply to the whole population of countries, and not just to the sample of, say, five countries for which data are at hand.

A set of hypothetical subjective competence scores for each of the five countries is displayed in table 22. Since these five countries are now seen as a random collection of countries selected from the larger population of countries, we are no longer as interested as before in the specific effect of each country on subjective competence. Rather, the five effects which can be computed, by taking the difference between the country means and the overall mean, form a random sample from a larger set of effects. In other words, these five values have been sampled from a random variable that would have provided different values if units of observation (countries) other than the five had been selected. Because of this, a model II type of analysis is also called a "random effect" model as opposed to a "fixed effect" model.

We are interested in the question of whether or not the country one lives in has an effect on one's subjective competence. This implies a null hypothesis which states that all the country effects are equal to zero. Analysis of variance offers a test of this hypothesis by determining whether or not the variance of all the effects is equal to zero. The only way a variance can equal zero is for all the values (when taken as deviations from their mean) to equal zero themselves.

Formal model and computations. The formal structures of model I and model II look very similar within the one-way domain, and the computations are identical. But a random model, in addition, permits the esti-

TABLE 22

Hypothetical Data for an Example of a One-Way Analysis of Variance, Model II, and the Resulting Analysis of Variance Table

		Country		
Great Britain	Italy	Mexico	United States	West Germany
8	5	7	9	6
7	4	6	8	5
6	3	5	7	4
5	2	4	6	3

Mean	6.5	3.5	5.5	7.5	4.5

Overall mean = 5.5

Source	Sum of squares	Degrees of freedom	Mean squares	F-ratio	Significance
Between countries	40.00	4	10.00	6.00	0.004
Within countries	25.00	15	1.67		
Total	65.00	19			

$\hat{\sigma}^2 = 1.67$ $\hat{\sigma}_a^2 = 2.08$

mation of the new concept of the variance of all the effects of a variable in the population.

The formal model states that y_{ij}, the j-th observation of the i-th group, is equal to the overall mean, plus the effect of being in the i-th group, plus the effect of all the other variables. This is expressed in the equation

$$y_{ij} = \mu + a_i + \epsilon_{ij}$$

where a_i is the effect of the explanatory variable in the i-th group, here country, and ϵ_{ij} is the effect on this particular observation of all other variables.

The model, furthermore, specifies that all the population values of the a's form a normal distribution with mean zero and variance σ_a^2. Similarly, the residual variable is assumed to have a normal distribution with mean zero and variance σ^2. A random model like the one we are dealing with here also requires the same number of observations in each group; a requirement not encountered in a model I type of analysis. In the example in table 22 each group has $n=4$ observations.

The hypothesis we want to test is that country has no effect on the dependent variable. In order for country to have no effect, all the a's have to equal zero, implying a variance equal to zero as well. The hypothesis of σ_a^2 equal to zero can be tested by the same F-ratio used for the model I type of analysis in chapter 2. We first find the total sum of squares by taking the difference between each of the 20 observations and the overall mean of 5.5, squaring all the differences, and adding up all the squares. That is

$$TSS = \Sigma(y_{ij} - \bar{y})^2$$

$$= (8-5.5)^2 + (7-5.5)^2 + \ldots + (3-5.5)^2 = 65.00$$

Similarly, the between groups sum of squares is found by taking the difference between each group mean and the overall mean, squaring all the differences, multiplying each square by the number of observations in each group, and adding up all the products. That is

$$BSS = \Sigma n(\bar{y}_i - \bar{y})^2$$

$$= 4(6.5-5.5)^2 + \ldots + 4(4.5-5.5)^2 = 40.00$$

Finally, the within groups sum of squares is found by taking the difference of the total sum of squares and the between groups sum of squares, here WSS = 65.00 − 40.00 = 25.00.

The degrees of freedom for the between sum of squares equals one less the number of groups, that is four degrees of freedom with five groups. The within groups sum of squares has degrees of freedom equal to the number of groups times the number of observations in each group minus one. With five groups and four observations in each group, the number of degrees of freedom becomes five times three, or 15.

By dividing each sum of squares by its degrees of freedom we get the corresponding mean square, as usual. For a random effect model it can be shown that the between group mean square, here equal to 10.00, is an estimate of the quantity $\sigma^2 + n\sigma_a^2$, σ^2 being the residual variance and σ_a^2 the variance of the population values, a. Furthermore, it can be shown that the residual mean square, here equal to 1.67, is an estimate of σ^2. Armed with these two pieces of information we can zero in on σ_a^2. The division of the between groups mean square by the residual mean square yields an estimate of the quantity $1.00 + n\sigma_a^2/\sigma^2$. If σ_a^2 really is equal to zero, the ratio calculated from our sample data should lie in the neighborhood of 1.00. On the other hand, if σ_a^2 is really not equal to zero, the observed ratio ought to be larger than 1.00. Turning to our example, we find the observed ratio equals F = 10.00/1.67 = 6.00, which is a good deal larger than 1.00. In order to decide whether the observed ratio is large enough to reject the hypothesis of a zero variance, we consult a table for the F-distribution. Such a table indicates that when the hypothesis of a zero variance is true the probability equals 0.05 of observing an F-value equal to or larger than 3.06, with 4 and 15 degtees of freedom. Since the F-value computed here is larger than 3.06, the hypothesis of a zero variance is rejected. In other words, country does have an effect on subjective competence, and that goes for the population of countries, not just for the five countries of the sample.

Estimating variance components. The rejection of the null hypothesis has established that the explanatory variable has an effect on the dependent variable. The next step consists of measuring, in some way, how large this effect is.

One such measure is provided by the share of the between sum of squares of the total sum of squares, which in our example is 40.00/65.00 = 0.615. The country variable, therefore, explains 61.5% of the total variation of subjective competence. Another, and related, way of measuirng the effect of the country variable would be to estimate the quantity a for each of the five countries in our sample. One could estimate the a-values for a country by taking the difference between the mean for that country and the overall mean, that is, $\hat{a}_1 = 6.5-5.5 = 1.0$, $\hat{a}_2 = 3.5-5.5 = -2.0$, and so on.

One difficulty with these numbers is that they do not tell us how large the a-values for all the other countries are. Moreover, the mean of such estimated a-values is by definition zero and, therefore, cannot give us a measure of the size of the effect. Instead we turn to the variance of the a's; the larger the variance, the larger the average a-value.

A good estimate of this variance (σ_a^2) can be found with the aid of results presented above. We know that

$$10.00 \simeq \sigma^2 + 4\sigma_a^2 \quad \text{and} \quad 1.67 \simeq \sigma^2$$

where the $\simeq$ sign indicates that the right hand expression is only estimated by, and not exactly equal to, the quantity on the left. Referring to the estimates (^) we can write:

$$\hat{\sigma}^2 = 1.67 \qquad \hat{\sigma}_a^2 = (10.00 - 1.67)/4 = 2.08$$

Since the variance of the a's exceeds the variance of the residuals in this case, the a-values on average are larger than the residuals. Another way of comparing these quantities is to take their square roots and obtain the standard deviations, giving us a measure of the average a-value and e-value (residual). With standard deviations of 1.44 and 1.29 respectively, we note that the average country effect is equal to 1.44. Recalling the large F-value we obtained for these data, we conclude that this effect is large enough to reject the hypothesis that the true population effect is equal to zero.

Overall, in one-way analysis of variance the random and the fixed models differ little with regard to computation and only slightly more so with regard to interpretation of results. In the two-way domain, however, these differences are far more pronounced.

Two Explanatory Variables

Formal model and computations. A random (type II) model of analysis of variance with two explanatory variables is based on the same idea as the random model with only one explanatory variable. The categories on both variables to be examined are samples from much larger populations of categories, and the aim of the analysis is to assess the effects of the explanatory variables on the dependent variable.

The example used earlier in this chapter for the one-way analysis of variance is extended in order to illustrate a two-way application. In addition to a sample of countries we assume that we have drawn a sample from the age list, supposing that an individual's age has an effect on his subjective competence. Assume that two ages had been picked, for example, the 30-year olds and the 60-year olds. With these two age cohorts and five countries we get 10 combinations of age and country. In the following example a sample of two respondents is drawn within each of these 10 combinations; the data are shown in table 23.

The formal model for this analysis specifies that the k-th observation in the cell defined by the i-th row and the j-th column, y_{ijk}, can be written as the sum

$$y_{ijk} = \mu + a_i + b_j + c_{ij} + \epsilon_{ijk}$$

The various terms in this sum have very much the same interpretations as the corresponding terms in the two-way analysis in chapter 3, which is a model I (fixed) type of analysis. The parameter μ is a constant, which can

TABLE 23

Hypothetical Data for a Two-Way Analysis of Variance, Model II, and the Resulting Analysis of Variance Table: Subjective Competence as Dependent Variable

| | | Respondent's Country | | | | | |
		Germany	Italy	Mexico	UK	US	Mean
Respondent's Age	30 years	7	6	3	8	9	5.9
		6	4	2	6	8	
	60 years	5	5	5	7	7	5.1
		4	3	4	5	6	
	Mean	5.5	4.5	3.5	6.5	7.5	5.5

Source	Sum of squares	Degrees of freedom	Mean squares	F-ratio	Significance
Age	3.2	1	3.20	1.18	0.30
Country	40.0	4	10.00	3.70	0.04
Interaction	10.8	4	2.70	2.45	0.09
Residual	11.0	10	1.10		
Total	65.0	19			

F (age)   = 3.20/2.70 = 1.18
F (country)   = 10.00/2.70 = 3.70
F (interaction) = 2.70/1.10 = 2.45

be estimated by the mean of all the observations. The next term, a_i, refers to the effect of the i-th category of the row variable (age). The effects of the two age cohorts are denoted as a_1 and a_2. These two quantities have been sampled from a large population of a-values, and it is assumed that the distribution of a-values in the population is a normal one with mean zero and variance σ_a^2. In order to assess the effect of age on subjective competence, we are not as concerned with the two values a_1 and a_2 as we are with the whole population of a-values. One way to specify that age has no effect is to specify that the population variance equals zero, which implies that all the a-values in the population are equal to zero.

Similarly, the term b_j refers to the effect of the j-th category of the column variable (country). Five columns give rise to five effects denoted as b_1, b_2, b_3, b_4, b_5. It is assumed, again, that the population from which these values come has a normal distribution with a mean of zero and a variance of σ_b^2. The null hypothesis that the country variable has no effect implies that the variance of the b's is equal to zero.

The interaction effect, produced by the combination of the i-th row and the j-th column, is captured by the term c_{ij}. With 10 combinations there are 10 interaction terms. Again it is assumed that they have been sampled from a population of interaction terms with a normal distribution having a mean of zero and a variance of σ_c^2. If the interaction has no effect, all the c-values must equal zero, and that again implies a variance equal to zero. Thus, the null hypothesis of no interaction effect can be written as $\sigma_c^2 = 0$.

The last term in the model, ϵ_{ijk}, specifies the effect on subjective competence stemming from all other variables. This is the residual variable, and it is assumed that the 20 residual terms in our example come from a population of residual terms which have a normal distribution with a mean of zero and a variance of σ^2.

The five sums of squares, degrees of freedom, and mean squares are computed the same way for a model II as for a model I analysis. The results pertaining to our example are shown in the analysis of variance table in table 23. The total sum of squares in the same as in the one-way analysis (table 22), and the country sum of squares is also unchanged. But the old residual sum of squares of 25.00 on 15 degrees of freedom has visibly declined as a result of having the age and interaction sums of squares extracted from it. It now stands only at 11.00 on 10 degrees of freedom.

The major difference between a model I and a model II type of analysis lies in the computation of the F-ratios. Recall that for the model I type of analysis (chapter 3) the F-ratios for the row variable, the column variable, and the interaction variable were each obtained by dividing the corresponding mean squares by the mean square for the residual variable. But this is no longer the case with a model II type of analysis. Here the mean square for the row variable (age) must be divided by the mean square for interaction in order to yield the F-ratio for the row variable. Similarly, the F-ratio for the column variable (country) is obtained by dividing the column mean square by the interaction mean square. Why is the interaction mean square chosen in place of the residual mean square for the F-ratio? The reason is that in the random model, unlike the fixed model, the mean squares for both column and row variables estimate quantities which include σ_c^2, the interaction variance. This is so because in a random model the interaction terms do not necessarily sum to zero across rows or columns—only across the whole table. Therefore, the model II algorithm for the F-ratios of row and column variables requires the interaction mean square in the denominator.

In our example the F-ratio for age (row variable) turns out $3.20/2.70 = 1.18$, whereas the F-ratio for country (column variable) turns out $10.00/2.70 = 3.70$. The F-ratio for interaction, which is computed just the same way as for model I, by dividing the interaction mean square by the residual mean square, is $2.70/1.10 = 2.45$.

It usually makes quite a difference whether a set of data is analyzed according to a random or a fixed model, even though it should always be clear from the way the data were collected which way the data should be analyzed. If the age and country mean squares of our example had been divided by the residual mean square, as required by model I, instead of by the interaction mean square, we would have obtained F-values of 2.91 and 9.09 respectively, rather than values of 1.18 and 3.70 as given in table 23. The effects would have proven far more significant under the fixed model than under the random model.

The F-values obtained under the random model prove insignificant for the age as well as the interaction variable. Neither the hypothesis σ_a^2 nor the hypothesis σ_c^2 can be rejected. Age differences have no effect on subjective competence. As for the country variable, the hypothesis σ_b^2 can be rejected. National differences retain their effect on subjective competence. Having performed these three tests, we are left with the task of estimating the size of each of the effects.

Estimating variance components. According to the statistical theory underlying the random model, the various mean squares and variances are related in the following way:

The row mean square is an estimate of	$\sigma^2 + n\sigma_c^2 + nc\sigma_a^2$
The column mean square is an estimate of	$\sigma^2 + n\sigma_c^2 + nr\sigma_b^2$
The interaction mean square is an estimate of	$\sigma^2 + n\sigma_c^2$
The residual mean square is an estimate of	σ^2

where n is the number of observations in each cell, r the number of rows, and c the number of columns in the table. These are the expressions which enabled us to compute the F-ratios the way we did, and they also lead us to estimates of the four variances in the model.

Taking the observed mean squares and replacing the theoretical variances by their estimates, using the letter s for estimates, we can write

$$3.2 = s^2 + 2s_c^2 + 10s_a^2$$
$$10.0 = s^2 + 2s_c^2 + 4s_b^2$$
$$2.7 = s^2 + 2s_c^2$$
$$1.1 = s^2$$

These equations can be solved successively. The variance of the residuals is estimated by the value 1.10 from the last equation above. The standard deviation of the residuals then equals 1.05. The next to last equation is then solved for s_c^2, the variance of the interaction effects.

$$s_c^2 = (2.7 - 1.1)/2 = 0.8 \qquad s_c = 0.89$$

Having determined the values for s^2 and s_c^2, we are now in a position to solve the first and second equation for s_a^2 and s_b^2 respectively

$$s_a^2 = (3.2 - 2.7)/10 = 0.05 \qquad s_a = 0.22$$
$$s_b^2 = (10.0 - 2.7)/4 = 1.82 \qquad s_b = 1.35$$

The results confirm our impression based on the F-values. The country variable emerges with the largest variance, the average country effect being equal to 1.35 (s_b), whereas the average age effect is indicated by a low 0.22 (s_a), well below the residual effect capturing all other variables which equals 1.05 (s). There remains, however, a noticeable interaction effect between age and country which shows an average of 0.89 (s_c). Comparing these components of variance or their corresponding standard deviations has helped us specify the effects of presumed explanatory variables on the dependent variable—subjective competence. This is a service of analysis of variance which works well when the explanatory variables are uncorrelated and when interaction is minimal.

5. OTHER MODELS

Three Explanatory Variables

In theory the analysis presented above can be extended to more than two nominal predictors, but with more variables one has to keep track of an increasing number of effects. Suppose three variables A, B, and C are included in the analysis. Besides three main effects, one for each variable, there are several interaction effects. Each pair of variables gives rise to an interaction effect, resulting in AB, AC, and BC interaction effects. Moreover, one interaction effect emerges from the combination of all three variables taken together, leading to a total of four interaction terms.

As with the simpler analysis involving two variables, the analysis of more variables requires the explanatory variables to be unrelated if the effect of each variable is to be separated from the other effects. The simplest way to keep explanatory variables from being related is to assign the same number of observations to each cell of the design. This procedure is feasible in experimental studies, but not in the usual survey study. As a result, in the nonexperimental domain of research, analysis of variance with many explanatory variables faces mounting difficulties.

Latin Square Design

Basic idea. As the investigator considers an increasing number of explanatory variables, he is faced with the problem of how to observe behavior in increasingly complex circumstances. In order to make the data gathering less cumbersome and less costly, and perhaps at all feasible, the investigator can avail himself/herself of several designs, one of which is called the "latin square" design.

Assume that the effects of three variables denoted A, B, and C on variable Y are to be examined; each of the explanatory variables having the same number of categories, say four, that is, A_1, A_2, A_3, and A_4, and similarly for B and C. In a psychological perception experiment, for example, the color of an object may be one of the explanatory variables, its four categories being blue, yellow, red, and white. Taken together, the three variables define $(4)(4)(4) = 64$ different combinations of categories. In order to investigate whether these variables have any effects on the dependent variable Y, one could set up an experiment to be analyzed by a three-way analysis of variance. With 64 distinct combinations, a minimum of 64 observations would be required, at least one for every combination. Gathering observations under these conditions is expensive and cumbersome, making it perhpas too difficult to conduct such projects.

A latin square model provides a solution to this dilemma. Such a design permits the investigator to study the effects of A, B, and C on Y using only $(4)(4) = 16$ instead of 64 observation as a minimum for analysis. This reduction is facilitated by the assumption that all four interaction effects—AB, AC, BC, and ABC—are zero. The interaction terms, it must be noted, capture the lion share of degrees of freedom, with AB, AC, and BD each responsible for $(4-1)(4-1) = 9$ degrees of freedom in our example, and the ABC interaction responsible for $(4-1)(4-1)(4-1) = 27$ degrees of freedom, adding up to a total of 54 out of a grand total of 63, with 64 observations. No more than nine degrees of freedom are required for the effects of the three variables. Once we ignore the interaction effects we can proceed with a far smaller number of degrees of freedom and, thus, observations. If the assumption of no interaction effects is tenable, the use of a latin square design offers a great saving in time and effort to the investigator.

Such a saving is obtained by careful selection of combinations of the three explanatory variables for which the dependent variable Y is observed. One chooses combinations which are balanced in such a way that the effects of A, B, and C can actually be assessed.

Design and analysis. One possible set of combinations serving this purpose is the following:

Combinations of A, B, and C	Value of Y
$A_1 B_1 C_1$	y_{111}
$A_1 B_2 C_4$	y_{124}
$A_1 B_3 C_3$	y_{133}
$A_1 B_4 C_2$	y_{142}
$A_2 B_1 C_2$	y_{212}
$A_2 B_2 C_1$	y_{221}
$A_2 B_3 C_4$	y_{234}
$A_2 B_4 C_3$	y_{243}
$A_3 B_1 C_3$	y_{313}
$A_3 B_2 C_2$	y_{322}
$A_3 B_3 C_1$	y_{331}
$A_3 B_4 C_4$	y_{344}
$A_4 B_1 C_4$	y_{414}
$A_4 B_2 C_3$	y_{423}
$A_4 B_3 C_2$	y_{432}
$A_4 B_4 C_1$	y_{441}

This list of combinations is not as randomly chosen as it may seem. Instead, a system underlies the selection. Each category of A occurs once with each category of B and C, each category of B occurs once with each category of A and C, and the same holds for C as well.

Another way of displaying the combinations is shown in table 24. Variable A is placed into the rows of the table, variable B into the columns, and the cells of the table are reserved for C. A close look at the table will convince the reader that a particular category of C occurs but once in each row and each column of the table. This assures that each category of a variable pairs but once with each of the categories of the other variables. The bottom part of table 24 presents numerical values of the dependent variable for each of the 16 cells of the design, one observation per cell. We now turn to the question of how to assess the effects of the three variables A, B, and C on Y.

Noting that the 16 values of Y are not all alike, we capture the extent of this dispersion by computing the total sum of squares, TSS. Here we have:

$$TSS = \Sigma(y_{ijk} - \bar{y})^2 = 271.420$$

TABLE 24
Combinations of Categories Used in a Latin Square Design, with Data

	B_1	B_2	B_3	B_4
A_1	C_1	C_4	C_3	C_2
A_2	C_2	C_1	C_4	C_3
A_3	C_3	C_2	C_1	C_4
A_4	C_4	C_3	C_2	C_1

				Means
12.0	10.5	10.7	11.2	11.100
15.9	10.3	15.2	7.5	12.225
12.7	11.2	13.6	11.1	12.150
7.7	0.1	8.6	1.7	4.525
Means 12.075	8.025	12.025	7.875	10.000

with 15 degrees of freedom, since there are 16 observations. In order to find how much the three variables have contributed to this sum, we must locate the mean of Y for each of the categories of A, B, and C.

The four observations in the first row belong to the first category of variable A. Their mean is

$$\bar{y}_{1..} = (y_{111} + y_{124} + y_{133} + y_{142})/4$$

$$= (12.0 + 10.5 + 10.7 + 11.2)/4 = 11.1$$

The notation used here reserves the first subscript of Y for the categories of A, the second subscript for the categories of B, and the third subscript for the categories of C. Thus, the mean for A_1 is denoted as $\bar{y}_{1..}$, with the dots placed in the second and third subscripts indicating that we have summed over these subscripts. Similarly, we find

$$\bar{y}_{2..} = 12.225 \qquad \bar{y}_{3..} = 12.150 \qquad \bar{y}_{4..} = 4.525$$

The mean values of Y for the observations which go with A_1, A_2, A_3 and A_4 have turned out different, and the extent of these differences is summarized by taking the deviation of these means from the overall mean and adding up the squares of those deviations. With an overall mean of 10.0 we get

$$4(11.100 - 10.000)^2 + 4(12.225 - 10.000)^2$$

$$+ 4(12.150 - 10.000)^2 + 4(4.525 - 10.000)^2 = 163.035$$

Above, each of the squares has been multiplied by 4, which is the number of observations for each category of A. The sum of these squares is that part of the total sum of squares which is due to the presence of the variable A. If this variable had no effect on Y, the means of Y for the four categories of A would have been roughly equal. The sum given above measures how unequal the means are and, therefore, how much effect A has on Y. Below we also show how one can test the null hypothesis that the means are equal in the underlying population, which implies that the differences we have observed in our data result from chance fluctuations only.

The sum of squares for B is found in a similar way. The four means are different, and the magnitude of the differences is measured by subtracting each mean from the overall mean, $\bar{y} = 10.0$, squaring each difference, and multiplying each square by four, the number of observations in each category. The sum of squares for B is found by adding those numbers; that is

$$SSB = \Sigma 4(\bar{y}._{j}. - \bar{y})^2 = 67.290$$

The sum of squares for C is more cumbersome to obtain. In order to find the mean of the dependent variable for C_1, we collect the four observations located along the diagonal running from the upper left to the lower right of the table. This gives us $\bar{y}._{.1} = 9.4$. Similarly, for $\bar{y}._{.2}$, the mean for C_2, we collect those four values from the table which refer to C_2; and the same procedure is followed to obtain $\bar{y}._{.3}$ and $\bar{y}._{.4}$. The effect of variable C is measured the same way as that of A and B, and we get

$$SSC = 4\Sigma(\bar{y}._{.k} - \bar{y})^2 = 38.665$$

The residual sum of squares is found by subtracting the sums of squares for A, A and C from the total sum of squares. The various sums of squares with their degrees of freedom, mean squares, and F-ratios are shown in table 25.

A latin square design, as has been demonstrated, has allowed us to assess the effects of three variables by observing only a small fraction—16 of 64—of the possible combinations of the three explanatory variables. Such a procedure, it must be remembered, is only permissible if the interaction terms can be assumed to be insignificant. A violation of this assumption calls the use of this type of design into question (Winer, 1971: 514-538).

Nested Designs

A variety of designs for collecting and analyzing data exist beyond those discussed up to this point. From among the remaining designs we

TABLE 25
Analysis of Variance Table for a Latin Square Design

Source	Sum of squares	Degrees of freedom	Mean square	F-ratio	Significance
A	163.035	3	54.345	509.80	0.0000
B	67.290	3	22.430	55.15	0.0001
C	38.655	3	12.885	31.69	0.0004
Residual	2.440	6	0.407		
Total	271.420	15			

briefly take up the nested design, sometimes also called hierarchical design. In its most simple form the nested design is an incomplete two-way design. It is incomplete in the sense that not all the data are available which would allow for a two-way analysis the way such an analysis was discussed in chapters 3 and 4.

As an example of data calling for a nested design take congressional districts organized within each of the 50 states (that is, those states with more than one district). If one were to assess the effect of district versus the effect of state level forces on some dependent variable, one would construct a design in which the districts were nested within their respective state. A given district is located in only one state. Thus, the two explanatory variables—district and state—cannot be completely crossed the way country and education were crossed in chapter 3. Table 26 presents a fragment of the nested design, listing three states, S_1, S_2, S_3, with three congressional districts nested within them. Within each district, n individuals are observed with regard to the dependent variable Y. As for the notation, the score of the second individual in district 9 located in state 3 is denoted as Y_{392}, to use an example.

A nested design in conjunction with a random model of analysis was employed by Stokes (1965, 1967) in his study of electoral effects. As in the example above, congressional districts were nested within states. Districts, state, and the nation as a whole were observed with regard to voting turnout as well as the party division in congressional elections across time. The purpose of the analysis was to separate out the effects due to district, state, and national forces as they bear on voting turnout and the partisan division of the vote. Stokes presents the variance components for the three effects and compares their size across a long historical span, noting striking trends in the variation of these effects over time.

Analysis of Variance and Regression

A final note is devoted to the connection between analysis of variance and regression analysis. The latter topic is treated in more detail by Uslaner's paper in this series. At this point we merely intend to hint at the interface between the two types of analysis. A set of observations studied by analysis of variance can always be reanalyzed using regression methods with suitably constructed dummy variables. This is so because all these methods are special cases of the so-called general linear model. In this section no proof is given for this equivalence; rather we want to show, by way of an example, how a one-way analysis of variance can be used to complement a simple regression analysis.

TABLE 26
Form of the Data for a Nested Design

| State | Congressional District | | | | | | | | |
	c_1	c_2	c_3	c_4	c_5	c_6	c_7	c_8	c_9
s_1	y_{111} y_{112} . . . y_{11n}	y_{121} y_{122} . . . y_{12n}	y_{131} y_{132} . . . y_{13n}						
s_2				y_{241} y_{242} . . . y_{24n}	y_{251} y_{252} . . . y_{25n}	y_{261} y_{262} . . . y_{26n}			
s_3							y_{371} y_{372} . . . y_{37n}	y_{381} y_{382} . . . y_{38n}	y_{391} y_{392} . . . y_{39n}

Table 27 presents a set of hypothetical subjective competence scores classified by education. For the sake of simplicity, the education variable is scored from 1 to 5. The regression analysis of these data results in the line with equation

$$y' = 4.033 + 1.453x$$

TABLE 27
Hypothetical Subjective Competence Scores Classified by Education

Education				
1	2	3	4	5
7	8	9	9	11
6	7	8	10	13
5	6	7	11	
		6		

Regression analysis:

Source	Sum of squares	Degrees of freedom	Mean squares	F-ratio	Significance
Regression	52.354	1	52.354	33.95	0.00006
Residual	20.046	13	1.542		
Total	72.400	14			

Analysis of variance:

Source	Sum of squares	Degrees of freedom	Mean squares	F-ratio	Significance
Groups	59.400	4	14.850	11.42	0.001
Residual	13.000	10	1.300		•
Total	72.400				

Combined:

Source	Sum of squares	Degrees of freedom	Mean squares	F-ratio	Significance
Regression	52.354	1	52.354	40.27	0.0001
Deviation from line	7.046	3	2.3486	1.81	0.21
Residual	13.000	10	1.300		
Total	72.400	14			

and a correlation coefficient $r = 0.86$ and $r^2 = 0.75$. The F-value for the test that the slope equals zero in the underlying population turns out to be 33.95 on one and 13 degrees of freedom. The various sums of squares and degrees of freedom are shown in the analysis of variance table for the regression analysis in table 27.

The analysis above is done under the assumption that the relationship between X and Y is linear. One way to check the assumption of linearity is to do a one-way analysis of variance on the same data. The reason for this is that the residuals in regression analysis are measured as deviations from the regression line, whereas the residuals in analysis of variance are measured as deviations from the group means. If the regression line passes through all the group means, the residual sum of squares will turn out the same for the regression analysis as for analysis of variance. But if the relationship between X and Y is not linear, then the regression line will not pass through all group means, and as a result, the regression residual sum of squares will exceed the analysis of variance residual sum of squares.

By comparing the two sets of results, summarized in table 27, we find that the two residual sums of squares differ by as much as 7.046. This is the part of the total sum of squares that is due to the fact that the means do not lie on the regression line. In the bottom part of table 27 this sum is entered as "deviation from line" along with the residual sum of squares obtained in the analysis of variance and the regression sum of squares. The null hypothesis that the group means lie on the regression line is tested with the appropriate F-ratio. Since the value of this F-ratio (1.81) proves insignificant, we conclude that in the population from which the data are drawn the group means do not deviate from the regression line. The deviation noted in our sample results from chance fluctuation. This result confirms the assumption that the relationship between X and Y, indeed, is linear.

6. CONCLUSION

We entered into the discussion of the method known as analysis of varinace by way of a substantive example. The example was taken from the general theme of political participation viewed from a comparative perspective. We were interested in the questions of

(a) how strongly the five nations included in the Almond and Verba data set differed from each other with respect to "subjective competence" in politics, and

(b) whether national (cultural) differences were more crucial for the overall level of subjective competence than was a country's level of social and economic development.

To what extent have the types of analysis presented in this paper helped us answer these questions?

One-way analysis of variance employing a fixed model has supplied a test of the (null) hypothesis that national differences (between the five countries) have no effect on subjective competence. This analysis also furnished an estimate of the proportion of variation in subjective competence scores explained by differences between the five nations. The results showed that the effect of national differences, while highly significant, did not turn out very strong. With this finding in hand we probed further to determine the effect of social and economic development—tapped by the indicator "education"—in comparison with the effect of national differences. This called for an extension of the analysis into a two-way format.

Two-way analysis of variance enables the investigator to separate the effects of two nominal predictors, test for the significance of each of their effects, and estimate the proportion of overall variation due to each effect. Under conditions of experimental research—allowing for uncorrelated predictors—two-way analysis of variance handles these assignments without resistance. Yet outside this realm of research the comfort ends. Where predictors are correlated—as they are in most survey situations—the use of the two-way algorithm does not furnish an immediate answer. Now the investigator must accommodate the joint effect of the two predictors, aside from the separate effects of each of them as well as interaction. This complication severely limits the investigator's ability to resolve questions of the kind which ask about the strength of effects. The closer the relationship between the two predictors, the less can be said about their separate effects.

In our example, the overlap between the predictors was fairly substantial. The effect of national differences on subjective competence, as observed in the one-way analysis, was sharply reduced ty the introduction of development (education). Similarly, the effect of education on subjective competence dropped sharply in the presence of the country variable in two-way analysis. This common drop testifies to the relationship between country and development (education). Once this commonality is set aside, it seems that of the two predictors social and economic development ranks higher than national—that is, cultural—differences in terms of fostering a sense of subjective competence.

Having grappled with the intricacies of a nonorthogonal design, we took on what is known as the "random model." Unlike the fixed model, the random model is designed for situations where only a sample of categories instead of the full set of categories of the nominal predictor(s) is available; apart from the sample of respondents within such categories. The random model helps the analyst make inferences about *all* categories of that predictor on the basis of the few randomly sampled categories; an inference, for example, from the five nations in the Almond and Verba study—presumably randomly selected—to all nations, that is from differences between five nations to national differences as such.

Both the fixed and the random models, along with many other models of statistical analysis, make certain assumptions which we have noted on a number of occasions. These assumptions, however, need not always be met in the strictest sense of their meaning. Under certain conditions they can be violated without the F-test losing its theoretical justification.

The assumption of a normal distribution of residuals for each category or combination of categories (two-way analysis) is one of them. This assumption need not bother us greatly so long as the sample of observations for each category is relatively large. This goes for both the fixed and the random model, as far as it was presented, although departures from normality are to be taken more seriously in the latter cases; the random model, moreover, also requires the effects to be normally distributed. To secure a large sample size for each category or combinations of categories would seem an easier job for survey studies than for experiments.

A second assumption worth our attention concerns the variance of the residual term, σ^2. This variance is supposed to be constant for all categories of the nominal predictor or combinations of categories. Such an assumption can be explicitly tested, but even where it is found lacking the analysis is not necessarily doomed. This holds true, for example, in one-way analysis employing a fixed model so long as the number of observations is equal for all categories. Whereas the planning of experiments typically assures such equality, survey studies are less equipped to do so. While the latter have the edge in size they suffer from the disadvantage of uncontrolled distribution.

Of the two models of analysis, the random model, on the whole, proves more demanding in terms of assumptions and less tolerant of violations of these assumptions than the fixed model; in return, it allows for inferences which reach far beyond those furnished by the fixed model. It is like the difference between five countries and the whole world.

NOTES

1. The data utilized in this study were made available by the Inter-university Consortium for Political Research. The data were originally collected by G. A. Almond and S. Verba. Neither the original authors nor the Consortium bear any responsibility for the analyses or interpretations presented here.

2. Each respondent in the five-nation survey was assigned a subjective competence score on the basis of responses to six questions. Three of the questions related to actions against a national law, as discussed above. The other three items involved

identical questions concerned with actions against a local regulation. A factor analysis provided the weights which were used to calculate a single score from those six items. The subjective competence score of respondent i, denoted S_i, was computed from the following equation:

$$S_i = 0.21Z_{1i} + 0.22Z_{2i} + 0.22Z_{3i} + 0.22Z_{4i} + 0.23Z_{5i} + 0.24Z_{6i}$$

The coefficients in this equation represent the loadings of the variables on the first factor of a principal components analysis divided by the eigenvalue of that factor (sum of squares of loadings). The z-notation indicates that each variable was standardized to have a mean of zero and a variance of one. The six variables are

Z_1: Action against bad local regulation

Z_2: Could change bad regulation

Z_3: Would act on bad regulation

Z_4: Action against bad national law

Z_5: Could change bad law

Z_6: Would act on bad law

The subscript i on each of the variables indicate we are dealing with the score for the i-th respondent.

3. For an exposition of this view see Norman H. Nie, G. Bingham Powell, Jr. and Kenneth Prewitt (1969) "Social Structure and Political Participation: Development Relationship, Part I, in *American Political Science Review,* 63 (June): 361-378. Their comparative attempt to explain rates of political participation puts heavy stress on developmental over idiosyncratic national factors: "The effects of discrete national experiences will be minimal in comparison to the effects directly attributable to economic development" (1969: 371).

4. For a further discussion of the problem of multicollinearity in regression analysis, see Uslaner (1976, forthcoming).

REFERENCES

ALMOND, G. A. and S. VERBA (1963) The Civic Culture. Princeton: Princeton Univ. Press.

ANSCOMBE, F. J. and J. W. TUKEY (1963) "The examination and analysis of residuals." Technometrics 5, no. 2 (May): 141-160.

BOX, G.E.P. (1953) "Non-normality and tests on variance." Biometrika, vol. 40: 318-335.

COCHRAN, W. G. (1965) "The planning of observational studies of human populations." J. of the Royal Statistical Society, series A, 128, part 2: 234-265.

――― and G. M. COX (1957) Experimental Designs. New York; Wiley.

DUNN, O. J. and V. A. CLARK (1974) Applied Statistics: Analysis of Variance and Regression. New York: Wiley.

FRANCIS, I. (1973) "A comparison of several analysis of variance programs." J. of the American Statistical Assn. 68, no. 344 (December): 860-865.

HAYS, W. L. (1973) Statistics for the Social Sciences. Second ed. New York: Holt, Rinehart & Winston.

KEPPEL, G. (1973) Design and Analysis: A Researcher's Handbook. Englewood Cliffs, N.J.: Prentice Hall.

KIRK, R. E. (1968) Experimental Design: Procedures for the Behavioral Sciences. Monterey, Calif.: Brooks/Cole.

NIE, N. H., G. B. POWELL, Jr., and K. PREWITT (1969) "Social structure and political participation: developmental relationship, part I." American Pol. Sci. Rev. 63 (June): 361-378.

NIE, N. H. et al. (1975) Statistical Package for the Social Sciences. Second ed. New York: McGraw-HIll.

OSIRIS Manual (1973) Ann Arbor, Mich.: Inter-university Consortium for Political Research.

SCHEFFE, H. (1959) The Analysis of Variance. New York: Wiley.

SNEDECOR, G. W. and W. G. COCHRAN (1967) Statistical Methods. Ames: Iowa State Univ. Press.

STOKES, D. E. (1967) "Parties and the nationalization of electoral forces," pp. 182-202 in W. N. Chambers and W. D. Burnham (eds.) The American Party Systems. New York, London: Oxford Univ. Press.

––– (1965) "A variance components model of political effects," pp. 61-85 in John M. Claunch (ed.) Mathematical Applications in Political Science. Dallas: Arnold Foundations.

USLANER, E. (1976) Regression Analysis: Simultaneous Equation Estimation. (forthcoming) Sage University Papers in Quantitative Applications to the Social Sciences. Beverly Hills: Sage.

WINER, B. J. (1971) Statistical Principles in Experimental Design. New York: McGraw-Hill.

GUDMUND R. IVERSEN, *associate professor of statistics and a statistician at the Center for Social and Policy Studies, Swarthmore College, received his Ph.D. from Harvard University. His articles have appeared in serveral scholarly journals, including the* Public Opinion Quarterly, Psychometrika, World Politics *and the* American Journal of Sociology. *He is currently coauthoring a book on statistical analysis of individual and group data.*

HELMUT NORPOTH *is an associate professor of sociology at the Zentralarchiv für empirische Sozialforschung at the University of Cologne. He received his undergraduate education at the universities of Freiburg and Berlin, and his Ph.D. from the University of Michigan. He is currently researching electoral behavior and coalition formation in West Germany.*